Pedagogy of Geography

Practicals and Practicum

Books by Dr. Mousumi Boral

Pedagogy of Geography: Modern Theoretical Perspective
Pedagogy of Geography: Practicals and Practicum

Pedagogy of Geography

Practicals and Practicum

Mousumi Boral

M. Sc. (Geogr.), B.Ed. M.Ed.,
Ph. D. (Sc.), M. A. (Edu.)

Rs. 995

ISBN: 978-93-91978-59-4

2024
First Published in India

Pedagogy of Geography:
Practicals and Practicum

Published by:
SHIPRA PUBLICATIONS
LG 18-19, Pankaj Central Market
I.P. Ext., Patparganj, Delhi 110092, India
+91 11 47322068; 96500 28065, 9810522367
info@shiprapublication.com
www.shiprapublication.com

Foreword

In today's globalized world, geography is a vital subject that requires a comprehensive approach to teaching. The Pedagogy of Geography Teaching is an exceptional resource that covers both theoretical and practical aspects of the subject, offering teachers and learners a holistic view of geography. This book offers diverse teaching strategies that engage students and make learning geography more interactive and enjoyable, even for visually-impaired learners, making it an inclusive resource for all.

Additionally, Pedagogy of Geography Teaching equips students and teachers with the knowledge and skills necessary to understand and address contemporary issues effectively. With its emphasis on inclusivity and contemporary issues, this book is an essential resource for all geography educators. I strongly recommend it to anyone looking to improve their understanding and teaching of geography. The book's comprehensive coverage and focus on current trends and developments in the field make it an excellent choice for educators and learners alike.

Prof. Madhumita Bandyopadhyay
NIEPA
New Delhi

Preface

Throughout human history, most societies have sought to understand their place in the world, and the people and environments around them. Geography asks spatial questions—how and why things are distributed or arranged in particular ways on Earth's surface and how they develop and change over time. It looks at these different distributions and arrangements at many different scales. Geographers explore both the relationships between the physical properties of Earth's surface and the human societies spread across it and the varied cultural exchanges.

Whether geography is thought of as a discipline or as a basic feature of our world, developing an understanding of the subject is vehemently important. Thereby the approach to the study of geography has changed over several decades. Thinking geographically helps people to be aware and remain connected among and between places and important events.

This book makes an effort to broaden the teaching ambiance with relevant inclusions of the matter and also attempts to provide additional information to interested readers with comprehensive illustrations and justified analysis in simple and lucid language.

I would like to convey my heartfelt thanks to Priyanka Patra my geography student who has significantly helped me with her dedicated contributions. It is worth mentioning my younger son Shridipta who had been a constant companion in the making of this book. I owe him immensely for his relentless encouragement. Acknowledging the moral support and belief in my work, was invaluable inspiration received from my elder son Shrimanta.

Needless to say, I am grateful to all persons who have contributed their utmost to make it a success. It is hoped that the users will enjoy studying the book as much as the author has enjoyed writing it. Comments and suggestions for further improvement of the book are always welcome.

Kolkata
26 January, 2024

Mousumi Boral

Contents

PRACTICALS

1

Identification of Rocks

The scientific study of rocks is called *Petrology* which is an essential component of Geology. The subject matter consists of the origin, association, occurrences, mineral composition, chemical composition, texture, structure, and physical properties of rocks. *Petrography* deals with the descriptive part of the rocks and *Petrogeny* deals with the mode of formation of rocks.

Geologists study rocks because they contain clues about what the Earth was like in the past. A historical record of a planet can be discovered. Moreover, the past environment, pre-historic plant and animal species and climate can be studied. Moreover, topography can also be interpreted and the historical study of soil formation can be done.

Definition of a Rock

A rock is a naturally formed, non-living earth material consisting of minerals and other particulate matter. Primarily the three types of rocks are *Igneous, Sedimentary* and *Metamorphic*. Each type can be further classified into different types according to their mode of formation, mineral content, colour, structure, texture and fossil content.

Economic Significance of Rocks

1. It provides nutrients needed for plant growth.
2. Rocks are a storehouse of minerals.
3. They contain valuable deposits of coal, oil and natural gas
4. They provide building stones for the construction of roads, dams and other structures.
5. Rocks get weathered and ultimately lead to the formation of soil.

Definition of a Mineral

Whereas, minerals are defined as natural, inorganic solid substance which has a crystalline structure with a particular chemical composition. Whereas, the rock is composed of several minerals, and is generally classified based on the process of its formation.

Identification of Rocks

1. Colour of the rock is the effect of the composition
2. Texture is the result of crystals and grains present
3. Reaction to acids and alkalis

The other essentials needed for identifying rocks are hand-lens or magnifying glass, dilute hydrochloric acid, a nail, a knife and surely a streak plate. These enable one to view the presence of fossils, rock layers or foliations, glassy look, pores and interlocking crystals in the rocks.

Igneous Rocks

Igneous rocks have been formed by the solidification and cooling of magma. They may have crystals but have no stratification or fossils. Examples- Granite, Basalt and Dolerite.

Granite: Granites are coarse-grained, light-coloured, acidic and holocrystalline plutonic igneous rocks with a medium-specific gravity (2.63 to 2.75). They are hard and compact rocks composed essentially of quartz, feldspar and mica, usually with some ferromagnesian minerals. Feldspar is identified by its lath shape in which cleavage may be present, while quartz is identified by its vitreous lustre and mica by its flakes mafic. Both plagioclase and mafic minerals tend to form subhedral crystals while intergranular spaces are occupied by anhedral quartz. This typical interlocking of grains forms the usual granitic texture. Sometimes there is some micrographic intergrowth of quartz and feldspar. Some of these are porphyritic. Reading the larger-scaled structures, some have a bonded or layered structure. High-level granitic rocks may develop a drusy structure.

Basalt: Basalt makes up over 98 per cent of all volcanic rocks. It is a very fine-grained and dark-coloured mafic igneous rock with a typical volcanic texture, very high specific gravity and high hardness. It is very compact and is characterized by conchoidal fractures. Individual minerals cannot be identified by the naked eye. The characteristic minerals are calcic plagioclase, pyroxene and iron ore with or without olivine. They may be vesicular or the vesicles may be filled with secondary minerals forming amygdales and thereby giving rise to the amygdaloidal structure. The basaltic magma which erupted underwater forms pillow-like masses giving rise to basaltic pillow lava. The individual pillows have a glassy crust and a more or less crystalline interior which may show radiating contraction cracks.

Dolerite: Dolerites are the hypabyssal representatives of the gabbros and basalts, finer-grained than the former but coarser-grained than the latter. They occur mostly as dykes and sills and their texture is frequently ophitic or sub-ophitic. They are dark-coloured mafic igneous rocks with high specific gravity (2.64 to 3.12), high hardness and high compactness. Feldspar laths are randomly oriented forming criss-cross arrangement-the interspaces are filled up by ferromagnesian minerals in which traces of cleavage can be seen. As it is prone to weathering, local patches of reddish-brown tint can be often distinguished.

Sedimentary Rocks

Sedimentary rocks have been formed by the processes of deposition, cementation and solidification. They do not contain crystals and have stratifications and fossils. Examples-, Sandstone, Limestone and Conglomerate.

Sandstone: Sandstone is very fine to very coarse-grained, light-coloured, moderately compact and moderately hard sedimentary rocks with a rough feel and medium specific gravity. It is dominantly composed of detrital sand. Grains pack together against the grain to form a framework that provides that primary solid support in the deposit. The framework is cemented together in a matrix of ferruginous, siliceous, argillaceous and calcareous sediments. The texture is typically clastic-granular. Graded or cross-bedded structures may be present. Depending on grain provenance their main types are recognized as quartzose sandstone, arkose or feldspathic sandstone and lithic sandstone.

Limestone: Limestone is a rock whose primary constituents are composed dominantly of aragonite and calcite. They are very fine-grained, light-coloured with moderate specific gravity and very compact but of low hardness (easily scratched by a knife or even fingernail). The texture is typically non-clastic calcarenite, calcarenite and calcilutite. It strongly reacts with cold and dilute hydrochloric acid. A salient feature is that most of the limestones are particulate and bimodal-ie. they consist initially of relatively large carbonate particles mixed in various proportions with carbonate mud. Very often compositional layering characterized by different colour bands may be present.

Conglomerate: Conglomerates are clastic sedimentary rocks of rudaceous type. These are pebbly rocks of medium or fine size (over 2 mm in diameter). These are variegated in colour with moderate to high specific gravity and moderate compactness. Being mostly of fluvial, glacial and shoreline origin, these consist of typical mud-free gravels or their lithified equivalents. The texture is typically clastic as lithic clasts are the essential constituents. The pebbles, cobbles, gravels rock fragments, and granules are cemented together in a fine-grained matrix of ground mass of ferruginous (reddish brown), siliceous (light coloured), in a fine-grained matrix of ground argillaceous (muddy odour) and calcareous (reaction with dilute hydrochloric acid) sediments.

Metamorphic Rocks

Metamorphic rocks have been formed by the metamorphism processes wherein all igneous and sedimentary rocks undergo tremendous changes due to intense heat and pressure. They may be crystalline, foliated or non-foliated but possesses no fossils. Example- Marble, Quartzite and Schist.

Marble: Marbles are coarse-grained, light-coloured rocks with high compactness, low hardness and medium specific gravity. They are produced by the metamorphism of limestones and are, hence, dominantly composed of calcite. They strongly effervesce with dilute hydrochloric acid. Generally, they show an even-grained texture. Some show a crude foliation due to the segregation of minor silicate constituents or in some cases parallel alignment of lustrous flakes of graphite while some bear the imprint of stain in the form of lensoid outlines.

Quartzite: Quartzites are very hard granular rocks produced by the recrystallization of sandstone. They are light-coloured, fine to medium-grained rocks with high compactness, low to medium specific gravity, granoblastic texture and an overall vitreous lustre without any cleavage. They are non-foliated and composed largely of quartz.

Schist: Schists are most often associated with regional metamorphism in medium-grade metamorphic environments. They are medium-grade metamorphic, medium-grained, light-coloured rocks with low hardness, low compactness and medium specific gravity. They are dominantly composed of flaky minerals (mica) which are oriented in a preferred direction, imparting foliations in the rock. The grains of quartz and feldspar are interspaced with the foliation. The texture is typically schistose formed due to the parallel arrangement of flaky minerals. They break readily into parallel-sided slabs, due to the presence of phyllosilicates and the parallel alignment of amphibole prisms and fibres.

Rock Cycle of Transformation

The rock cycle is an idealized cycle of processes undergone by rocks in the earth's crust, involving igneous, sedimentary and metamorphic rocks interrelated with each other.

The Rock Cycle describes the transformations of rocks and subsequently the formation of new topographic structures on the dynamic Earth.

Identification of Rocks		
Igneous Rocks		
 Granite	 Basalt	 Dolerite
Sedimentary Rocks		
 Sandstone	 Limestone	 Conglomerate
Metamorphic Rocks		
 Marble	 Quartzite	 Schist

<table>
<tr><th colspan="6">Identification of Minerals</th></tr>
<tr><td colspan="2">
Quartz</td><td colspan="2">
Feldspar</td><td colspan="2">
Mica</td></tr>
<tr><td colspan="3">

Calcite</td><td colspan="3">
Pyroxene</td></tr>
<tr><td colspan="3">
Galena</td><td colspan="3">
Olivine</td></tr>
<tr><td colspan="3">
Hematite</td><td colspan="3">

Gypsum</td></tr>
</table>

2

Scale

The scale is the ratio between map distance and ground distance. Hence the map is simply the scale model of the real world. Map scales are represented in three basis and reversible forms—a statement, a numeric ratio and a graph and accordingly they are respectively called Statement scale, Ratio scale and Geographical scale. The scale is normally placed within a base, being associated with the north arrow positioned exactly middle, just above the scale bar. The location of the scale box depends on the page layout. The best cartographic presentation is decided by page borders, location, size, and shape of the outline map, available space, map index, key, or legends.

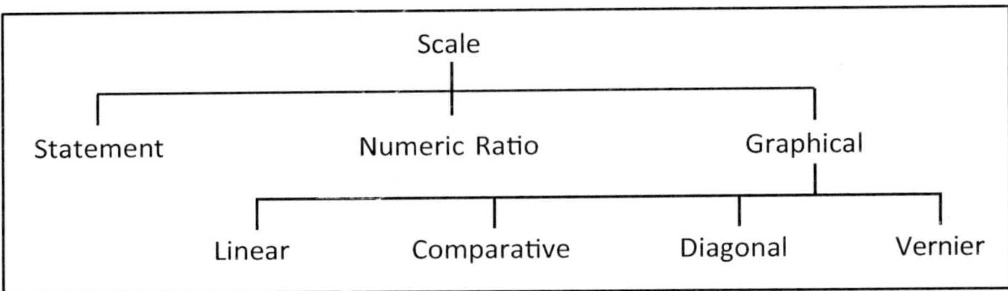

Statement Scale

A map scale may be in form of a simple statement in which the map distance is always expressed as a unit length. This is known as a statement scale and is written on a map as, 1 cm to 25 km or 1 inch to 50 miles. The value on the left-hand side of the statement indicates the map distance (cm/inch) and that on the right-hand side indicates the ground distance (km/mile).

Merits

(i) It is simplest to express.
(ii) It is not necessary to draw the scale like linear or graphical in the map.

Demerits

(i) A layman, ignorant of the various scale of measurements will not be able to read the map.
(ii) A reproduction of the original map by reduction or enlargement necessitates a determination of the scale all over again.

Numeric Ratio Scale

A map scale may also be expressed as a numeric ratio in which the numerator is the map distance and the denominator is the corresponding ground distance, both being expressed in the same unit of measurement. Therefore, it is a dimensionless fraction. It is also known as a ratio scale or representative fraction (R.F.), *e.g.* 1: 100000.

Merits

(i) The greatest advantage of a numeric ratio scale is that a map with a given R.F. can be used universally.

(ii) A statement scale or graphical scale in any system of measurement can easily be computed from R.F. and cartographically plotted as well.
Thus, 1: 100000 may be written as 1 cm to 100000 cm (CGS system)

$$\text{or, } 1 \text{ cm to} \frac{100000}{100000} \text{km}$$
$$\text{or, } 1 \text{ cm to } 1 \text{ km}$$

Demerits

(i) The scale will be changed by enlarging or reducing the map.

(ii) The previous knowledge about the unit of measurement is necessary to understand the representative fraction.

Graphical Scale

A map scale may also be cartographically represented by a linear graph or line. This is known as the graphical scale. This may be of four different forms—1. Linear Scale, 2. Comparative Scale, 3. Diagonal Scale, 4. Vernier Scale.

Linear Scale

This is the simplest form of graphical scale and is shown as a linear graph. The primary fundamental division (at least three) represents the ground distance in multiples of 1, 5 or 10. The primary division on the extreme left is subdivided into a suitable number of secondary divisions. The length of each secondary division depends on the fractional length up to which precision in measurement is desired.

Merits

(i) A linear scale helps to determine easily the idea of distance between two places on a map by visual observation or by a simple scale of measurement.

(ii) The ground distance can be easily measured from the map distance.

(iii) Enlargement or reduction does not change the measurement of the scale of the map as the scale also be reduced or enlarged consequently. This merit is not available in the case of the R.F. scale.

Demerits

(i) Drawing of a linear scale requires much time and may involve defects.

(ii) The respective unit of scale may not be acceptable all over.

(iii) Change in the unit of measurement of the scale may involve less of accuracy.

Problems

Example 1: Construct a linear scale to show 4 km in the primary division and 1 km in the secondary division and the given R.F. is 1: 200000.

Step 1: We know, 100000 cm = 1 km

$$\text{or , } 1 \text{ cm} = \frac{1}{10000} \text{km}$$

$$200000 \text{ cm} = \frac{1 \times 200000}{100000} \text{ km} = 2 \text{ km}$$

Step 2: The value of each primary division is 4 km.
∴ 2 km on the ground distance represented by 1 cm on the map distance.

1 km on the ground distance represented by $\frac{1}{2}$ cm on the map distance.

4 km on the ground distance represented by $\frac{1 \times 4}{2}$ cm on the map distance.
= 2 cm on the map distance.

Step 3: For the convenience of drawing the total length of scale will be—
2 cm on the map represented by 4 km on the ground.
2 × 7 cm on the map represented by 4 × 7 km on the ground.
14 cm on the map represented by 28 km on the ground.

Step 4: The value of each secondary division = 1 km.

$$\therefore \text{Total number of primary division} = \frac{\textit{Total ground length of scale}}{\textit{Ground length of primary division}}$$

$$= \frac{28\,km}{4} = 7 \text{ divisions.}$$

$$\therefore \text{Total number of secondary division} = \frac{\textit{Total ground length of each primary division}}{\textit{Ground length of each secondary division}}$$

$$= \frac{4\,km}{1} = 4 \text{ divisions.}$$

Methodology

- The map distance for each primary division is first evaluated from either statement or R.F.
- Select and fix the magnitude of ground distance for one primary division and secondary division.
- Based on the given problem or R.F., compute the map distance in cm or inch for the corresponding value of primary division.
- Draw a horizontal straight line.
- Divide the straight line with the help of a divider, diagonal scale, and perpendicular straight lines are drawn at each division with the help of set square.
- The first primary division from the left-side is drawn, then graphically sub-divided into secondary division with the desired precision.
- The drawing is then properly labelled and bordered.

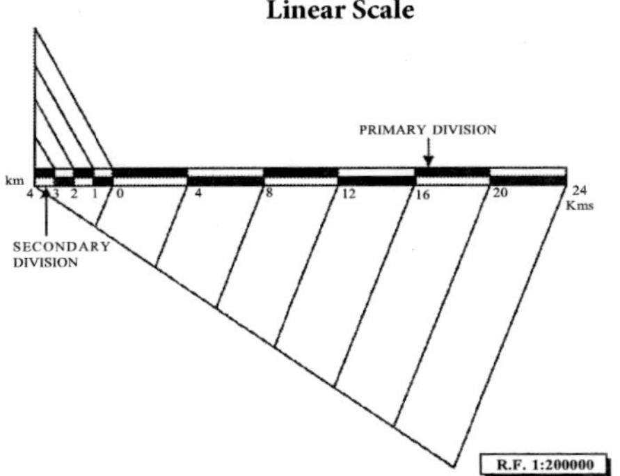

Figure 1: *A linear scale showing divisions in kilometres and metres*

A linear scale has been drawn on the basis of above calculation.

Example 2: Draw a linear scale to show 2 km in the primary division and 500 m in thesecondary division with R.F. = 1 : 98000.

Solution: R.F. = 1 : 98,000

$$\therefore \text{1cm of scale represents 98,000 cm}$$

$$\text{or, } \frac{98,000}{100,000} \text{ km } [\because 100,000 \text{ cm} = 1 \text{ km}]$$

$$\equiv 0.98 \text{ km}$$

$$\text{or, } 0.98 \text{ km} = 1\text{cm}$$

$$\therefore 15 \text{ km} = \frac{15 \times 0.98}{100,000} = 14.7 \text{ km}$$

Since 14.7 km is not a round figure, take 15 km as a round figure and calculate the length of the line as:

14.7 km are represented by 15 cm

$$\text{1 km is represented by } \frac{15}{14.7}$$

$$\text{10 km is represented by } \frac{15 \times 10}{14.7} = 10.20 \text{ cm}$$

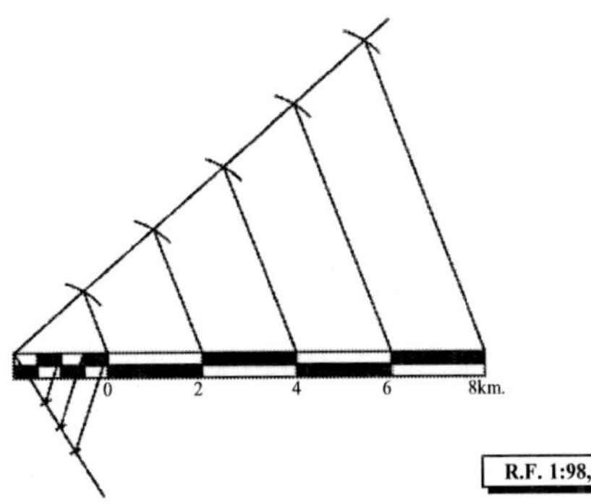

Figure 2: *A linear scale showing divisions in kilometres and metres*

Now a line of 10.20 cm is drawn which represents 10 km. It is divided into five primary divisions of 2 km each and then the first primary division is divided into four secondary divisions of 500 m each.

Example 3: Draw a linear scale to show 10 km in the primary division and 2 km in the secondary division when the given R.F. is 1 : 411,842.

Solution: R.F. = 1 : 411,842

$\therefore$ 1cm of scale represents 411,842 cm

$$\text{or, } \frac{411,842}{100,000} \text{ km } [\because 100,000 \text{ cm} = 1 \text{ km}]$$

$$\equiv 4.11 \text{ km}$$

$\because$ 4.11 km are represented by 1 cm

$\therefore$ 1 km is represented by $\dfrac{1}{4.11}$

$\therefore$ 50 km is represented by $\dfrac{1}{4.11} \times 50$ cm = 12.16 cm

Now a line of 12.16 cm is drawn which represents 50 km. It is divided into five primary divisions of 10 km each and then the first primary division is divided into five parts to show 2 km each. The given scale is drawn in the following section.

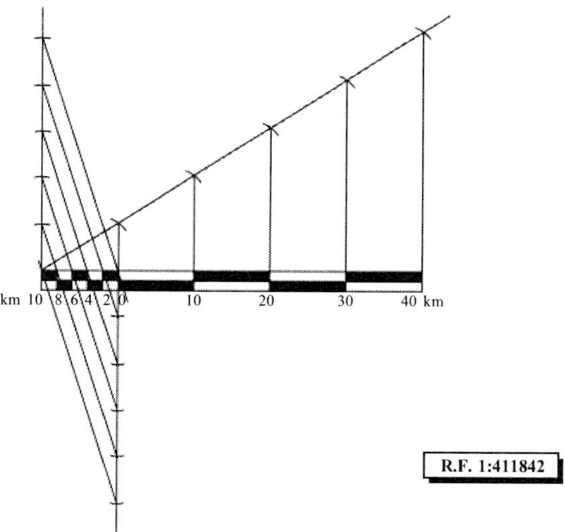

Figure 3: *A linear scale showing divisions in kilometres*

Example 4: Draw a linear scale to show 40 km and 10 km as primary and secondary divisions respectively when the R.F. is 1 : 2,000,000.

Solution: R.F. = 1 : 2,000,000

$\therefore$ 1cm of scale represents 2,000,000 cm

$$\text{or, } \frac{2,000,000}{100,000} \text{ km } [\because 100,000 \text{ cm} = 1 \text{ km}]$$

$$\equiv 20 \text{ km}$$

$\because$ 20 km is represented by 1 cm

$\therefore$ 1 km is represented by $\frac{1}{20}$ cm

$\therefore$160 km is represented by $\frac{1}{20} \times 160$ cm = 8 cm

Now a line of 8 cm is drawn which represents 160 km. It is divided into four primary divisions of 40 km each and then the first primary division is divided into four parts to show 10 km each as secondary division. The given scale is drawn in the following section.

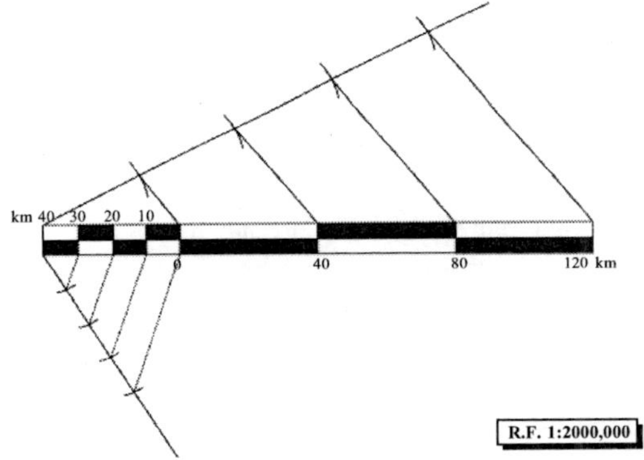

Figure 4: *A linear scale showing divisions in kilometres*

3

Temperature and Rainfall Graphs

A temperature and rainfall diagram is a composite diagram that represents the variation of temperature and rainfall of different months of a place in a year. Temperature can be diagrammatically represented with the help of a line graph (usually in red ink) in degrees Centigrade or Fahrenheit. The highest and lowest temperatures and the total rainfall in a year are depicted in the graph and thereby the climatic characteristics of the concerned place can be studied. The rainfall is usually graphically depicted as shaded columns (bar graphs usually in blue ink) for each month of a year like that of the temperature in cm/mm or in inches.

Meteorological stations, scattered all over the globe, are constantly gathering a wide range of data on temperature and rainfall for the construction of weather maps or temperature graphs. Climatic data mainly include the amount of rainfall collected through rain gauge and temperature which is measured with the help of various meteorologically designed thermometers.

Temperature is recorded daily which is laterally converted to the mean monthly temperature of a single month (according to the number of days of that month). The difference between the hottest and the coldest month is given as the annual range of temperature.

In the case of *rainfall*, daily records are added at the end of the month for the monthly rainfall amount. These monthly rainfalls are added at the end of the year (for 12 months) to find the annual rainfall and this annual amount can be divided by 12 months to obtain the mean annual.

Since the plotting of rainfall and temperature of a place for a year gives the learner, a fair idea about the weather of that particular place. This type of simple climate graph displaying yearly temperature and precipitation statistics for a particular location enables the learner to have an idea of the place. Most geographers worldwide have chosen the temperature-rainfall graph to identify the climatic region to which that particular place belongs and then can ensure regional planning.

Equatorial Climate

The equatorial region is hot and wet throughout the year. It is also known for its dense evergreen forest cover. It is situated between 5° to 10° North and South of the Equator in both hemispheres. More than 200 cm of annual rainfall and constant high temperature are the common features in this region.

Table 1: *Colombo (Sri Lanka)- 06°54'N, 79°25'E; Elevation -7 m*

Month	*January*	*February*	*March*	*April*	*May*	*June*	*July*	*August*	*September*	*October*	*November*	*December*	*Mean/ Annual*
Temp°C	26.1	26.4	27.2	27.8	28.0	27.2	27.2	27.2	27.2	26.7	26.1	25.8	26.9
Rainfall in mm	89	69	147	231	371	223	135	109	160	348	315	147	2344

Table 2: *Madang (New Guinea)- 5°14'S, 145°45'E; Elevation- 6 m*

Month	*Janu-ary*	*Feb-ruary*	*March*	*April*	*May*	*June*	*July*	*Au-gust*	*Septem-ber*	*Octo-ber*	*No-vember*	*De-cember*	*Mean/ Annual*
Temp°C	27.2	26.9	26.9	27.2	27.2	27.2	27.2	27.2	27.2	27.5	27.5	27.5	27.2
Rainfall in mm	307	302	378	429	384	274	193	122	135	254	338	368	3484

Highest temperature: 28 ° C

Lowest temperature: 25.8 ° C

Highest Rainfall: 429 mm

Lowest Rainfall: 69 mm

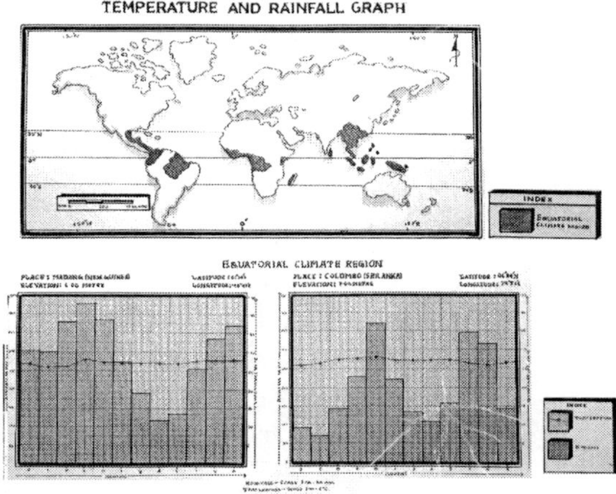

Interpretation

From the temperature values, it has been seen that there is a great uniformity of high temperatures throughout the year. There is hardly any variation. The annual range of temperature is not more than 30°C. In Colombo, the highest temperature is 28°C in the month of May. The lowest temperature found in the month of December is 25.8°C.

Precipitation is heavy and well distributed throughout the year. There are no months without rainfall. Instead, there are two periods of high rainfall (one is Apr/May and the other is Oct/Nov), which occur in the equatorial region. This double peak coincided almost with the equinoxes. The double peak is the characteristic of equatorial climates which are generally not found in other climatic regions. Hence it can be concluded that these areas are in the northern hemisphere of the Equatorial climate belt.

Tropical Monsoon Climate

Its latitudinal extension is 10° to 30° North and South of the Equator mostly near the Tropic of Cancer and the Tropic of Capricorn. Basically, extensive deciduous trees are found in the monsoon region and these trees shed their leaves during the dry season to conserve moisture.

Table 3: *Kolkata (India)- 22°34'N, 88°24'E; Elevation- 6 m*

Month	*Janu-ary*	*Feb-ruary*	*March*	*April*	*May*	*June*	*July*	*Au-gust*	*Septem-ber*	*Octo-ber*	*No-vember*	*De-cember*	*Mean/ Annual*
Temp°C	19.6	22.0	27.1	30.1	30.4	29.9	28.9	28.7	28.9	27.6	23.4	19.7	26.4
Rainfall in mm	12	28	34	50	134	290	331	334	253	127	27	4	1624

Table 4: *Papiti (Tahiti)- 17°32'S, 149°34'E; Elevation- 92 m*

Month	*January*	*February*	*March*	*April*	*May*	*June*	*July*	*August*	*September*	*October*	*November*	*December*	*Mean/ Annual*
Temp°C	29.6	26.9	26.9	26.9	25.8	25.3	25.3	25.0	25.3	25.8	26.4	26.7	26.3
Rainfall in mm	251	244	429	142	102	76	53	43	53	89	150	249	1881

Highest temperature: 30.4 ° C
Lowest temperature: 19.6 ° C

Highest Rainfall: 429 mm
Lowest Rainfall: 4 mm

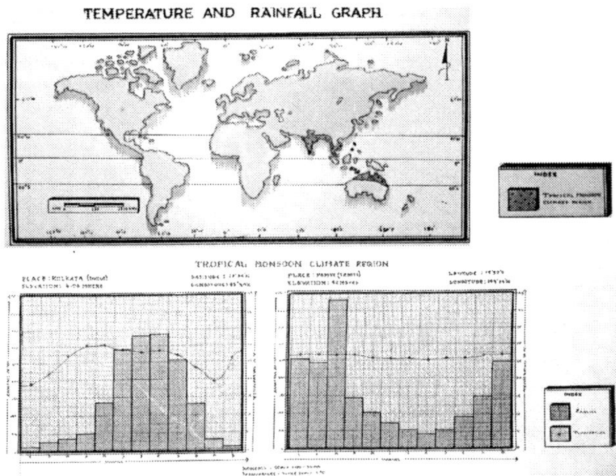

Interpretation

From the temperature values, it is observed that there is a great uniformity of temperature throughout the year in Papiti. But a small fluctuation in temperature has been observed in Kolkata in the year. The highest temperature has been observed in Kolkata during the month of May. It is because of the deficiency of rainfall that leads to the dry season which prevails during the months of April-May and December-January. Hence it is situated in the northern hemisphere in the ideal zone of Tropical Monsoon Climate.

The precipitation is varied all over the year in Kolkata. The highest amount of rainfall occurred in the month of July and August. But in Papiti, the rainfall is the same or more or less evenly distributed throughout the year with peaks in the December-January months. Situated at a 92-meter height and the highest rainfall recorded is 429 mm in March. But there is no month without rainfall.

From the above discussion, it can be interpreted that the uncertainty of rainfall and fluctuation of temperature have been observed throughout the year which are the main climatic characteristics of Tropical Monsoon Climate.

Hot Desert

In the tropics, these desertic regions are situated between 15° to 30° latitude in both hemispheres. Excess evaporation and less precipitation make these regions arid climatic zones. As a result, only the xerophytic vegetation thrives in these arid and semi-arid regions. These regions are mostly located in the western portions of the continents.

Table 5: *Aswan (United Arab Emirates, UAE)- 24°04'N, 32°53'E; Elevation- 116 m*

Month	*January*	*February*	*March*	*April*	*May*	*June*	*July*	*August*	*September*	*October*	*November*	*December*	*Mean/ Annual*
Temp°C	16.7	18.3	22.5	27.2	31.4	33.6	33.6	33.6	31.7	29.2	23.6	18.3	26.6
Rainfall in mm	2	3	3	3	3	3	0	0	0	2	2	2	23

Table 6: *Lima (Peru)- 12°05'S, 77°03'E; Elevation- 120 m*

Month	*January*	*February*	*March*	*April*	*May*	*June*	*July*	*August*	*September*	*October*	*November*	*December*	*Mean/ Annual*
Temp°C	23.3	23.9	23.6	24.7	19.4	17.2	16.7	16.1	16.9	18.1	19.4	21.1	20.03
Rainfall in mm	3	2	2	2	5	5	8	8	8	3	3	2	51

Highest temperature: 33.6 °C
Lowest temperature: 16.1 °C

Highest Rainfall: 8 mm
Lowest Rainfall: 0 mm

TEMPERATURE AND RAINFALL GRAPH

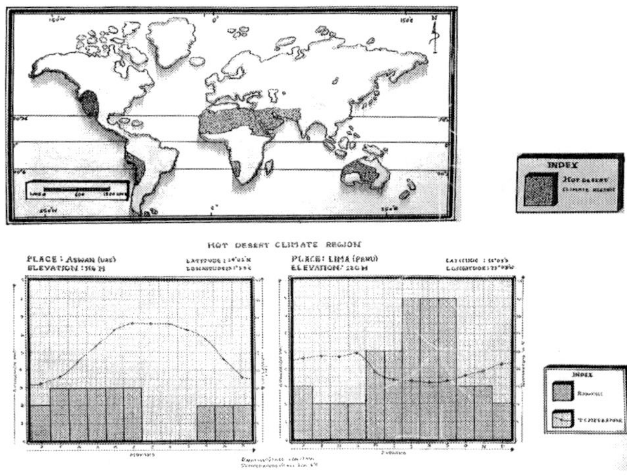

Interpretation

In Aswan, the climate is quite different from the climates of the equatorial or monsoonal regions. The highest temperature observed is 33.6°C and the lowest is 16.1°C. But the highest temperature was observed in the months of June, July, and August. It may be because of the negligible amount of rainfall. The amount of rainfall in these months is almost zero. In the other months, the amount of rainfall shows deficient trends. This is varied from 0 to 3 mm. Whereas Lima is situated in a hot climatic region so it receives rainfall with varying amounts from 2 mm to 8 mm in the year. The temperature remains almost consistent in all the months. But there is no month without rainfall.

From the above discussion, it can be concluded that the high temperature and negligible/low rainfall are the striking characteristics of the hot desert climatic regions.

Mediterranean Climate

It is situated between 30° to 45° N and S in the temperate climatic zone. This region bears a stark specialty than other regions with a distinct climatic and vegetation type. It mainly encompasses areas around the Mediterranean Sea but in the southern hemisphere, there are few regions that resemble such climatic types.

Table 7: *Marsais, (France)- 43°10′N, 05°23′E; Elevation- 79·6 m*

Month	*January*	*February*	*March*	*April*	*May*	*June*	*July*	*August*	*September*	*October*	*November*	*December*	*Mean/ Annual*
Temp°C	7.5	6.9	8.1	10.0	13.1	16.7	20.0	22.2	21.9	19.2	14.7	10.6	14.2
Rainfall in mm	48	38	46	51	48	25	15	23	66	94	79	56	589

Table 8: *Santiago (Chile)- 33°27′S, 70°42′W; Elevation- 522 m*

Month	*January*	*February*	*March*	*April*	*May*	*June*	*July*	*August*	*September*	*October*	*November*	*December*	*Mean/ Annual*
Temp°C	20.5	20.0	18.0	15.3	11.4	8.6	8.9	10.3	12.2	14.7	17.2	23.3	15.0
Rainfall in mm	3	3	5	13	63	84	76	56	30	15	8	5	361

Highest temperature: 23.3 ° C
Lowest temperature: 6.9 ° C

Highest Rainfall: 84 mm
Lowest Rainfall: 3 mm

TEMPERATURE AND RAINFALL GRAPH

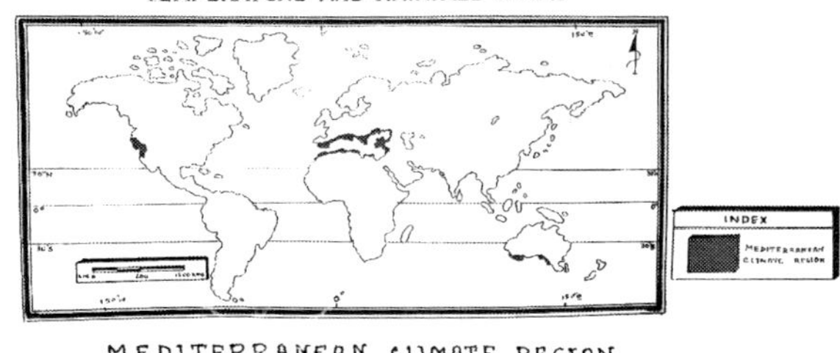

MEDITERRANEAN CLIMATE REGION

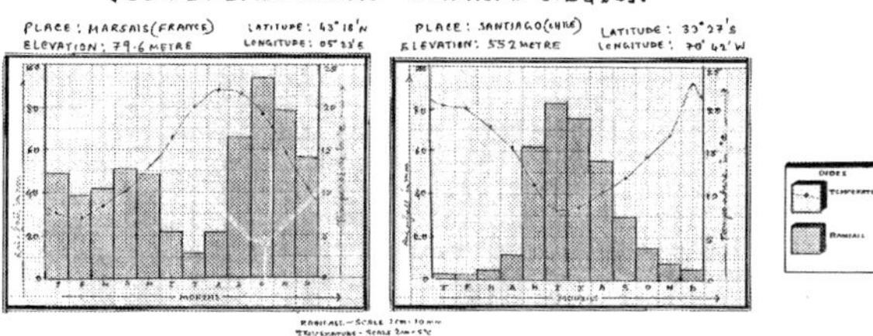

Interpretation

From the temperature values of Marsais (France), it is observed that the temperatures are highest in the months of August- September & lowest in January–February, and December, whereas, in Santiago (Chile), in the southern hemisphere lowest recorded temperature is in the June–July.

Whereas, the rainfall recordings are highest in the October–November, and December months and the lowest in July–August in Marsais in the Northern hemisphere. In the Southern Hemisphere, the opposite situation occurs, the highest rainfall is in June–July, and August months like in the Tropical Climatic region. But the stirring difference is that the temperature is lowest in these months. Moreover, rainfall is lowest in the month of January–February, and December.

So, from the above discussion, it can be said that the striking characteristics of winter rainfall primarily indicate the Mediterranean climatic type.

4

Drawing of Isolines

Isotherms, Isobars, Isohyets and Contours

Isolines

Isolines are lines of equal values [Iso = equal; lines = lines = isolines]. They mark the stage or state of uniformity. These lines of great importance are drawn or constructed to show the uniformity in the distribution of temperature (air), pressure (air), rainfall, relief (features) and other geographical aspects. Major isolines which are generally used are (i) isotherms, (ii) isobars, (iii) isohyets and (iv) contours.

Method of Drawing Isolines

Isolines are drawn by calculations based on specific points of interaction. The specific points are available from data collected from weather stations or survey offices. Weather offices supply data on air temperature, air pressure and rainfall of different stations or places. Specific spot heights or heights of the land are collected by means of land surveying or from survey offices. These data are plotted on a piece of paper or on maps. These maps or plotted sheets of paper are given to draw isolines.

The main problem in drawing isolines is finding out the location of the position of the said lines to their actual site. It can be achieved by following the *method of interpolation.*

Finding out the Position of Isoline

Interpolation is a method by which the position of a point is determined from the interaction of two other points located nearby.

Suppose, a point is to be determined between points A and B. The value of A is 80 and the value of B is 120. Now finding out the value of C to be located in between the two points. By the simple mathematical method, it can easily be determined (80 + 120) ÷ 2 = 100. Therefore, the value of C is 100. In the same manner, an iso-point can be easily determined.

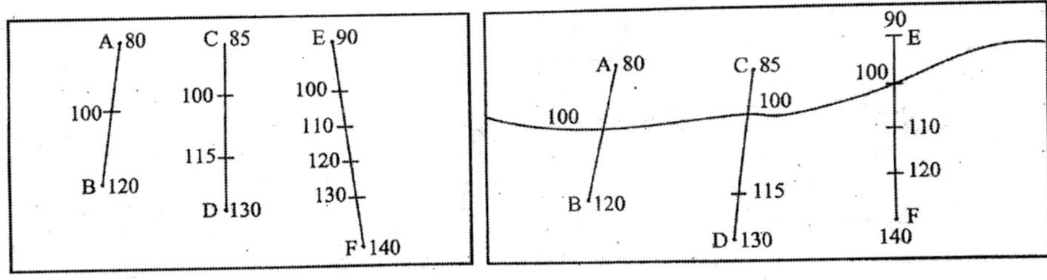

Figure 1: *Locating the exact iso-point (100) between two points of different values*

A graphical method of finding out iso-points and drawing out of isoline is shown in Figure 1.

Problem: To draw the given isolines on the practice paper with 50, 100 and 150 in Figure 2.

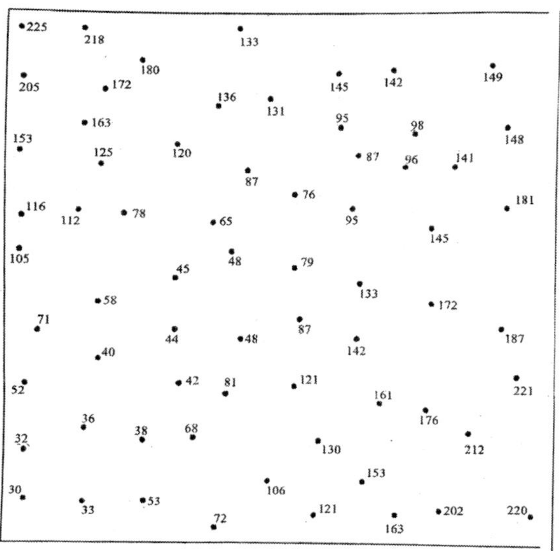

Figure 2: *Values to Draw Isolines (for practice)*

Isotherms

It is known that the distribution of temperature is not equal all over the earth's surface. The places which record equal temperatures in a particular season are joined by an imaginary line on the map which is known as the Isotherm. Isotherm means (Iso = equal, therm = temperature) equal temperature.

Definition

Isotherms are imaginary lines drawn on a given map joining places of the same average temperature for a specified period, supposing them to be at sea level (reduced to sea level).

Method of Drawing Isotherms

The following steps are:

(i) Conversion of temperature to the sea level location: All places on the earth's surface are not situated at the same sea level. Some places are adjacent to the coastal part, lying almost at sea level, while others are located at higher altitudes on mountains, hills and, plateaus. For their altitudes, differences in temperature occur. If isotherms are drawn according to the temperature of those places at different altitudes, a number of complications may arise for a simple interpretation.

To avoid such complications and eliminate the effects of altitudes, temperatures are reduced to sea-level temperature i.e., as if all the places are located at the same sea level. Thus, isotherms are drawn with the temperatures which are made reduced to sea level. At every 100 metres rise (ascent), the temperature reduces at the rate of 0.6°C (1°F at every 300 feet height).

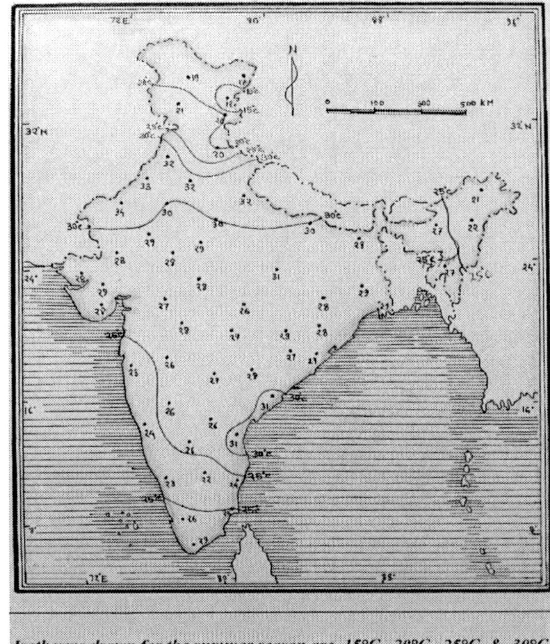

Isotherms drawn for the summer season are 15°C, 20°C, 25°C & 30°C

Map No 1: *Isotherms showing Summer Temperatures over India*

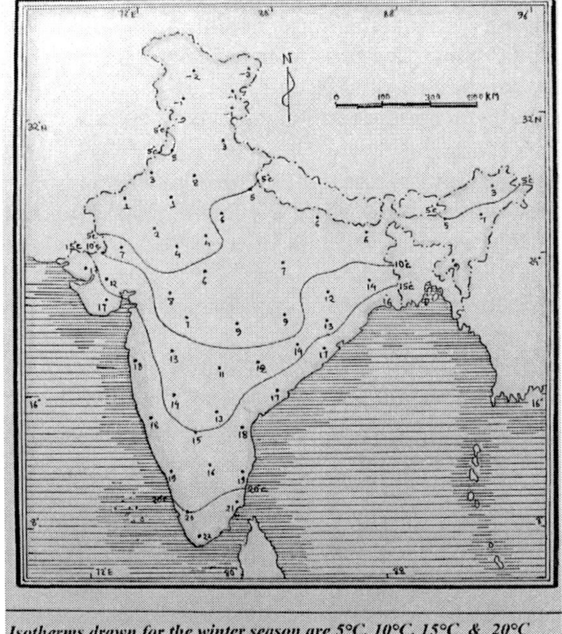

Isotherms drawn for the winter season are 5°C, 10°C, 15°C & 20°C

Map No 2: *Isotherms showing Winter Temperatures over India*

Accordingly, a reduction in temperature is to be made. If a place is at an altitude of 600 metres and its actual temperature is 20°C, its reduced temperature at sea level would be 20°C + 6°C = 26°C (because temperature decreases at the rate of 0.6°C for every 100 metres ascent).

(ii) Plotting the temperature values or data on a map: Thus, having prepared the temperature data reduced at sea level, students should proceed for plotting them on a map according to their placing of respective stations.

(iii) Drawing of isothermal lines: It is a simple act of joining the points of stations having equal temperatures shown on the map. In order to do this, the students need to follow the same method of drawing contour lines.

Precautions in Drawing Isotherm lines

The students need to:

1. Examine the location and distribution of the points, marking the values of temperature of the different places given on a map.
2. Join the places with a line having the same temperature (according to the method following for drawing contours).
3. Observe at first and then draw carefully so that a line does not pass through or cross any other line, having a different temperature.
4. Remember that an isotherm line never crosses another isotherm line.

Problem: Map No. 1 shows the summer temperature in India. Four isotherms are to be drawn at 5°C intervals like 15°C, 20°C, 25°C and 30°C.

Interpretation: These are quantitative areal maps where quantities are indicated by lines of equal value known by a multiplicity of such terms as isopleth (*iso* meaning equal and, *pleth* meaning lines). Wright (1944) proposed that isograms be used for all lines of quantity with two subdivisions- the isometric lines (*metron* meaning measurement) that represents a constant value or intensity pertaining to certain areas through which it passes.

Isotherms are lines which join places of equal temperature conditions. From the value of temperature of Map nos. 1 and 2 it becomes very clear for, the identification of summer maps or winter maps. The temperatures during summer are very high in contrast to the winter temperatures which are low. The most heated regions during the summer are north-west part of India in the states of Punjab, Rajasthan, and Delhi and more or less, all over central India, the temperature conditions remain quite high. By closely observing the isotherms drawn in the map of India showing winter conditions, it becomes clear that the lowest temperatures are recorded in the northern part of India in the states like Jammu and Kashmir, Delhi, Punjab, Rajasthan and, Haryana. The southern part of India remains comparatively warm because of the presence of the Bay of Bengal, Indian Ocean and the Arabian Sea, and also its southerly position in terms of latitudes.

Isobars

It is known that the distribution of pressure is not equal all over the earth's surface. The places on the map which record equal pressure in a particular season are joined by imaginary lines known as the Isobars. The instrument used to measure atmospheric pressure is called the barometer. Isobars mean (Iso-equal, bar-barometric pressure). These lines are on a weather map of constant barometric pressure drawn on a given reference surface. The isobaric pattern on a constant-height surface is extremely helpful in weather forecasting as temperature and wind are in close association with pressure conditions. Regions of low pressure at sea level tend to be areas of bad weather, especially in winter.

Definition

Isobars are imaginary lines drawn on a given weather map joining places of equal atmospheric pressure for a specified period, taking them to be at reduced sea level. A map with the isobar marked 1004 represents an area of high pressure, while the isobar marked 976 represents an area of low pressure.

Features of Isobars

1. The isobars are drawn in a similar manner to the isolines.
2. The numbers measure the atmospheric pressure in millibars.
3. Usually, isobars are drawn at intervals of two or four millibars (one thousandth of a bar).
4. The closer the isobars are together the pressure gradient is steeper and it is windier in the region.
5. If the lowest number is in the middle circle, this is a low pressure or depression which may render wind and rain.
6. If the highest number is in the middle circle, it is a high-pressure or anticyclone which results in dry and sunny weather conditions.

Problem: Map No. 3 shows the overall pressure conditions in India. Three isobars are to be drawn like 996 mb, 1000 mb, and 1004 mb.

Interpretation: The map depicts general atmospheric pressure conditions over Indian territory. The general trend of the isobars shows low-pressure conditions towards the interior locations and relatively high-pressure areas adjacent to the coast except in the

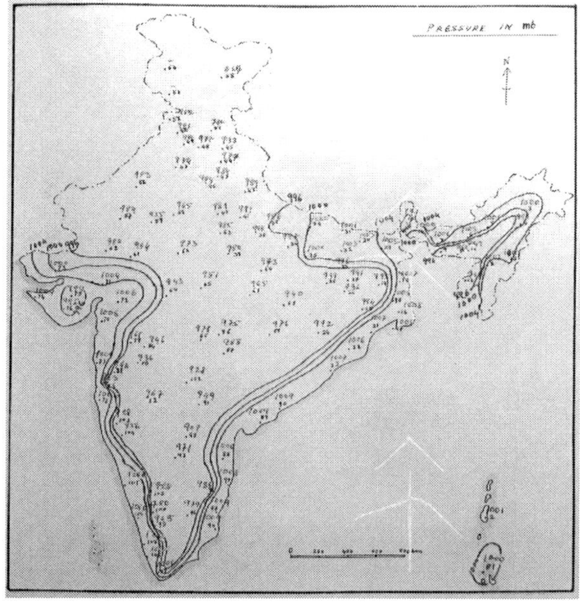

Map No 3: *Isobar Map of India*

north-eastern region where the wedge of high pressure is towards north-east. Moreover, the islands also exhibit an increasing trend of pressure. It exhibits the period which is at end of summer when low-pressure conditions exist on the interior of the landmass and high over the adjacent outer regions.

Isohyets

The isohyets maps are the quantitative maps which are indicated by the lines of equal value. The isohyets maps represent the constant value or intensity pertaining to certain area through which it passes. These isohyets are imaginary lines which join the places of equal rainfall area.

From the values of rainfall, it becomes very easier to the identification of distribution or trend of rainfall during summer or winter seasons. In these tropical regions, the rainfall is very high during the monsoon season and less during the winter season. The major rainfall occurs from mid-June to mid-September in West Bengal, Uttar Pradesh, Odisha, Punjab, Meghalaya, Maharashtra etc. Tamil Nadu region experiences rainfall twice a year i.e., summer monsoon (South-west monsoonal winds) and winter (north-east monsoonal winds) seasons. The amount of rainfall decreases from east to north-west ward. There are few rain shadow regions situated in India like one situated in south-eastern coastal India and another in the north-east region.

Problem: Map No. 4 shows the overall rainfall trend in India. Four isohyets are to be drawn like 60 cm,120 cm,180 cm, and 240 cm.

Interpretation: The map depicts the rainfall trend over the Indian subcontinent. The isohyets exhibit a decreasing trend of rainfall as one moves from the south, south-eastern regions towards the western and north-western regions. Moreover, the amount

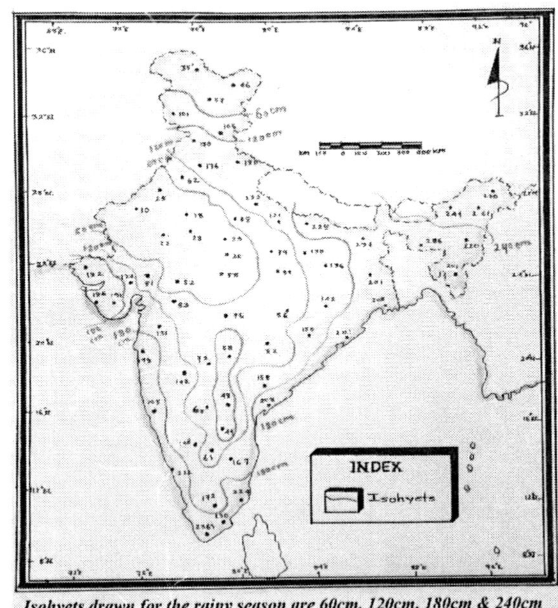

Isohyets drawn for the rainy season are 60cm, 120cm, 180cm & 240cm

Map No 4: *Isohyets showing Rainfall over India*

of rainfall is also low towards northern India as precipitation may be in the form of snow. Comparatively, high rainfall is recorded in north-eastern India. The isohyet map shows the temperature, pressure and wind conditions over the area.

Contours

Land areas on the earth's surface rise at different levels and heights above sea level. Hence, they have different altitudes. By their different elevations, they form different types of landforms, such as mountains, hills, valleys, plateaus, and plains. Land slope that rises from the sea level is known as altitude. In fact, altitude is the vertical distance above mean sea level, measured in feet or metres.

Definition

A contour may be defined as an imaginary line showing the same height or altitude on a map. Thus, the meandering lines which show the altitude of land on a map are called contours. On a map, land-heights are usually joined to draw a series of contour lines. A map showing the height of land by contours is called a *contour map*. Contour lines are lines showing the same height on the map. These are not straight lines, but are zigzag, winding or serpentine in nature. Values of their altitudes are marked within the line at definite intervals.

During surveying, spot heights are noted in the survey book, marking points with their respective heights (values). With the help of these heights, contours are drawn on a map according to certain intervals, such as 10, 20, 25, 50, 100, 200, 250 (metres or feet) and so on, as required or surveyed.

Method of Drawing Contours

1. Careful observation, examination and concentration of the mind in drawing contour lines at the definite interval (as directed) inf the location with spot heights are essential.
2. Drawing from the lower contour lines of the lowest spot heights towards the higher contours of the highest spot heights is required to avoid complications.
3. It should be noted that a contour never ends in the middle of a map. It starts from one end and then terminates at last on the other end or starts from one point and then goes on around the different directions and meets again, at last, forming a circular or elongated shape.
4. A contour never cuts away or crosses another contour. Sometimes, it may touch/ merge just another contour to show the steepest gradient as in the case of waterfalls or high/steep cliffs/mountains.
5. After drawing the contours, it is advisable to write their respective values in numerical form. Values should be given in writing either within the line or at the ends. In the case, of circular or semicircular contours, values should be given in writing within the lines.

The most careful issue in drawing a contour line is determining the location of points (in between spot heights) which are required but not given as spot heights or there is no marking of them. But in drawing, this line is needed to find out their position in between two spot heights. The points are to be found out from the positions of other spot heights. Thus, it is a matter of finding out their positions (Refer Figure 1)

According to the given map, a judicious choice of scale is to be made. To avoid miscalculations and misconceptions arbitrary markings need to be avoided. All the points are to be found out by proper calculation according to the given scale through the following method of interpolation of contour. In fact, interpolation is a method of finding out a position in between two poles *i.e.*, two points (spot heights) as in the case of contour drawing in which spot heights are given as points showing heights or altitudes. In determining or finding out a position of a point, according to the scale between two spot heights, the calculation should be made accurately and contours are then to be drawn through these points with the help of other spot heights.

By measuring the linear distance between the two points of given values, the spot value of the contour directed can be identified by calculations, placed and then drawn between them.

In the same way, a contour of 600 metres is to be drawn. In this case, at first, an examination of the places through which this line may enter is needed. The 600m contour line of this area (Figure 3) may proceed through a suitable point between 670 m and 550 m spot heights. The necessity is there to find out the 600 m. By minutely calculating the actual linear distance and distributing the difference (670-550) m on the line the exact point of drawing the contour of 600 m can be plotted.

***Example 1*:** Examine carefully how are contour lines of 100, 200, 300 and 400 metres drawn on the given diagram in Figure 3 and observe their positions.

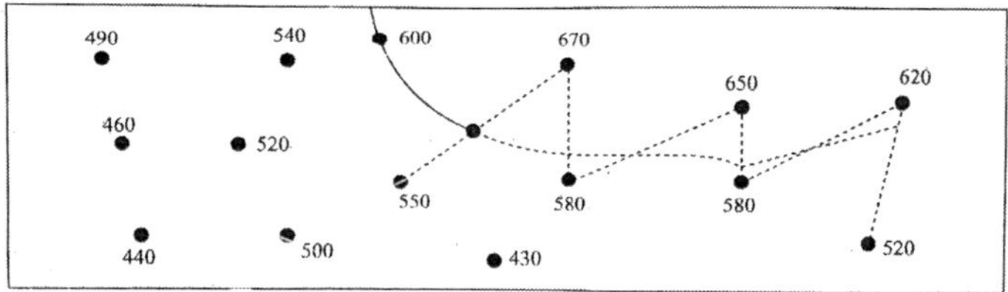

Figure 3.

Problem: To draw contours on the given sheet of paper with 100, 200, 300 and 400 metres in the given Figure 4.

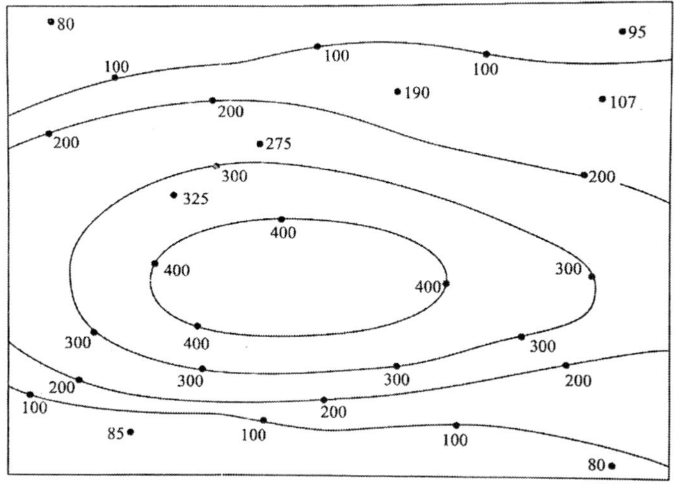

Figure 4.

5

Meteorological Instruments

To observe and record weather conditions and various elements of weather, different types of meteorological instruments are needed. The following are the weather instruments—

1. Six's Minimum and Maximum Thermometer
2. Hygrometer and
3. Fortin's Barometer.

Six's Minimum and Maximum Thermometer

- *Air Temperature Recording*: The warmth or coldness of the air is called air temperature. The sun is the source of all heat and energy in the atmosphere as well as in the earth. The atmosphere *i.e.* the air is not directly heated by the sun At first, the surface of the earth is heated up by the sun. Then the heat of the earth's surface is radiated to the layers of the atmosphere by the processes of conduction, convection and radiation.
- *Measurement of Temperature*: The temperature or the degree of the heat of air may be high (*i.e.* hot) or it can be very low (*i.e.* cold). It is measured with the help of an instrument known as Thermometer.

A Six's Minimum and Maximum Thermometer is employed to know the maximum and minimum temperatures of the day. A self-recording instrument known as a Thermograph is used for measuring temperature.

There are two common scales used for graduating a thermometer: [i] Centigrade Scale, [ii] Fahrenheit Scale.

The temperature of melting-point of ice is recorded as ± 0° in the Centigrade Scale and 32° in the Fahrenheit Scale. The temperature, at which water boils, is taken as 100° in the Centigrade Scale and 212° in the Fahrenheit Scale. It is, therefore, there are 100 divisions between the melting point of ice and the boiling point of water in the Centigrade Scale. Whereas in the Fahrenheit Scale there are 180 divisions between these two points.

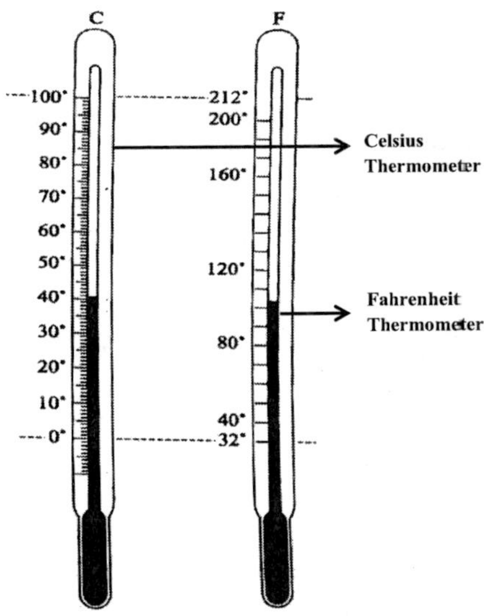

Figure 1: *Celsius and Fahrenheit Thermometer*

1. *Conversion Method from °F to °C Formula:*

   ∵ Temperature in °C = Temperature in (°F-32) × $\frac{5}{9}$

   ∴ If the temperature of a place is 90°F, the temperature in °C

   = Temp. (90-32) × $\frac{5}{9}$ = 32.2°C

2. *Conversion Method from °C to °F Formula:*

   ∵ Temperature in °F = Temperature in (°C × $\frac{9}{5}$) + 32

   ∴ If the temperature of a place is 17°C, the temperature in °F

   = Temp. (17 × $\frac{9}{5}$) + 32

   = 30.6 + 32 = 62.6°C

It is a thermometer designed to record both the maximum and minimum temperature of a day. The indexes can be made up and down with help of a small horse-shoe magnet to set them on the mercury surface.

At first, the thermometer has to be set before use. It can be done by moving each index up or down with the help of a magnet until they reach the mercury surface just in contact.

The reading at the lower ends of the indexes I_1 and I_2 register the minimum and maximum temperature of the day respectively. Then indexes are set again on the mercury surface by the magnet for reading of the next day.

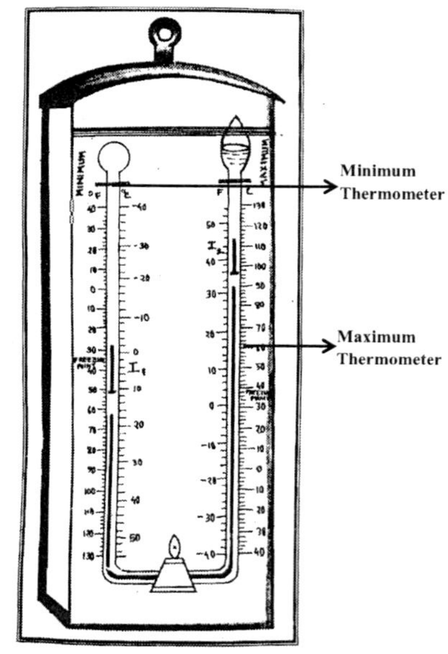

Figure 2: *Six's Minimum and Maximum Thermometer*

- *Temperature Recording Method*

The following table shows how to record temperatures (°C):

Table 1: *Diurnal Temperature Recording*

Date: 06.11.2022 and 07.11.2022 Place: Gandhinagar, Gujarat Time: 8.30 a.m.

Date	*Maximum Temperature*	*Minimum Temperature*	*Difference Temperature*	*Average Temperature*
05.11.2022	36°C	17°C	19°C	26.5°C
06.11.2022	36°C	18°C	18°C	27°C

Record taken by: India Meteorological Department, Meteorological Centre, Ahmedabad 382475

- *Terms of Temperature Data Calculation:*

Temperature varies from day to day and from time to time. The highest temperature of the day is called daily maximum temperature. The air gets maximum temperature in the afternoon at about 2 p.m. It is for the afternoon is hotter than noon-time. The lowest temperature of the day is called the daily minimum temperature. The minimum temperature is found at about 4 a.m. in the early morning, when the earth becomes cool due to rapid radiation.

1. *Mean daily temperature*: The average of maximum and minimum temperature recorded during a day is known as mean daily temperature. To get the mean daily temperature, the total of maximum temperature and minimum temperature is to be divided by 2.
   Example: On 5th November 2022 the maximum temperature of Gandhinagar was 36°C and minimum temperature was 17°C. Therefore the daily mean temperature on 5th November 2022 was:
   $$= \frac{\text{Maximum} + \text{Minimum}}{2} = \frac{36+17}{2} = 26.5°\text{C}$$
2. *Diurnal or daily range of temperature*: The difference between maximum and minimum temperature recorded in a particular day at a place is called Diurnal or Daily range of temperature.
   Example: If the maximum temperature of a place is 36°C and minimum temperature was 17°C, the diurnal or daily range of temperature is: (36° - 17°)C = 19°C
3. *Mean monthly temperature:* The average temperature of a month is known as mean monthly temperature. To get it, the total of daily temperature for all the days of a month is divided by the number of days of that month.
4. *Mean annual temperature*: The average of the monthly mean temperature of a year is known as the mean annual temperature. To get it, the total of mean monthly temperature of all the months of a year is divided by the number of months *i.e.* 12 months.
5. *Annual range of temperature*: The difference between the mean temperature of the hottest month and the mean temperature of the coldest month is called the mean annual range of temperature.
   Example*:* If the mean temperature of the hottest month is 35°C and the mean temperature of the coldest month is 17°C, the annual range of temperature is: (35°C - 17°C) = 18°C.

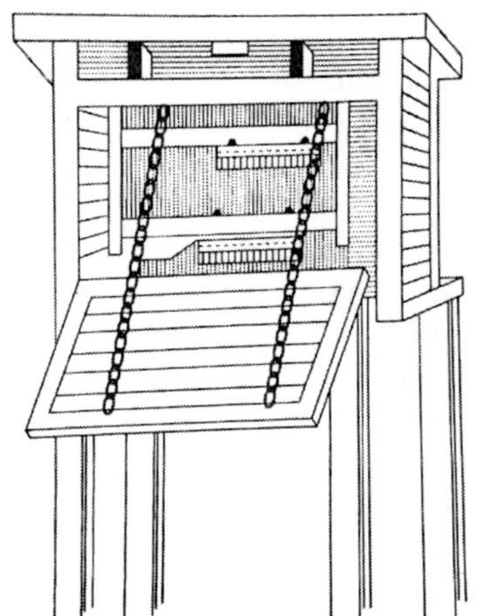

Figure 3: *Stevenson Screen*

Stevenson Screen

To obtain correct reading, the weather stations use the Stevenson Screen. It is manufactured in such a manner that the shade temperature of the air can be measured.

The Stevenson Screen consists of a wooden box, painted white and kept to stand at a height of about a metre placed away from building and trees. The four sides of the box are louvered to allow the free entry of the air. The roof has an intervening air space to prevent direct sunshine and radiant heat. One side is hinged so as to serve as a door. The instruments are kept inside the box supported on a frame. The instruments which are kept in the box are: 1. The Maximum and Minimum Thermometer, 2. Ordinary Temperature Thermometer and 3. Dry and Wet Bulb Thermometer (Hygrometer) etc.

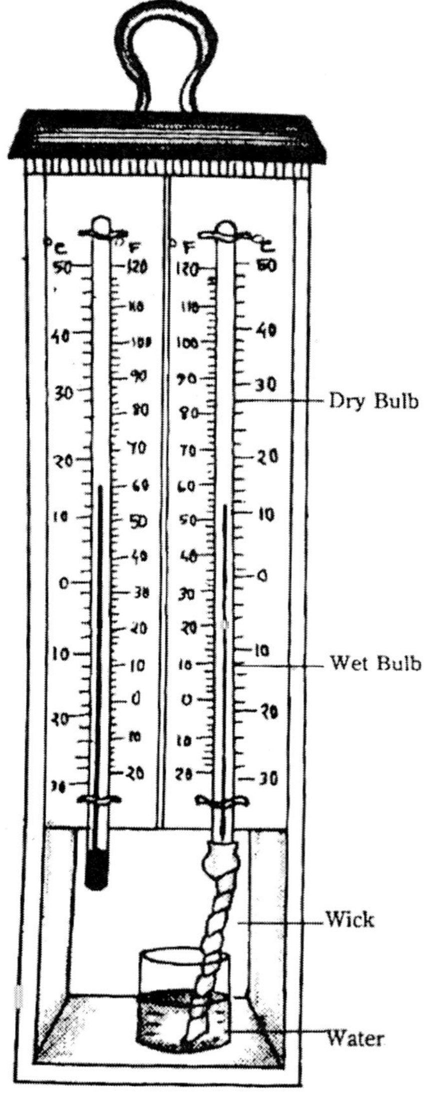

Figure 4: *Dry and Wet Bulb Thermometer*

Hygrometer or Dry and Wet Bulb Thermometer

To measure the humidity of air, an instrument called Hygrometer is employed. It consists of two thermometers: a dry bulb thermometer and a wet bulb thermometer. The wet bulb thermometer is covered with a piece of muslin or cotton cloth. It is kept constantly moist by a wick dipping into a small vessel containing water.

The dry bulb thermometer records the temperature of the atmosphere. The wet bulb thermometer measures a lower temperature due to cooling effect caused by the evaporating of water from the vessel. The rate of evaporation depends upon the relative humidity of the surrounding air and for this, fluctuation in temperature occurs in the wet bulb reading.

If the air remains saturated, evaporation would be little and the thermometer will record almost the same reading. When the relative humidity will be low, more evaporation will occur and the difference in temperature between the two thermometers will be high. (i) a small difference in temperature indicates high relative humidity; (ii) a large difference denotes low relative humidity and (iii) same reading in the two thermometers gives the indication of saturation. Actual amount of relative humidity is calculated by the difference of temperature from Standardise Table.

A table is given here to show how the amount of Relative Humidity varies at different temperatures:

Table 2: *Relative Humidity Table*

Dry Bulb Reading	*Depression of Wet Bulb (°C)*																		
(°C)	*0.5*	*1.0*	*1.5*	*2.0*	*2.5*	*3.0*	*3.5*	*4.0*	*4.5*	*5.0*	*5.5*	*6.0*	*6.5*	*7.0*	*8.0*	*9.0*	*10.0*	*11.0*	*12.0*
-5	85	71	58	44	30	17	4												
4	87	73	61	48	35	53	12												
3	88	75	64	52	40	29	18	7											
2	89	77	67	56	45	34	24	14	3										
1	90	79	69	59	49	39	30	20	10	1									
0	90	81	71	61	52	44	34	25	16	7									
+1	90	81	73	64	55	47	38	29	20	13	4								
2	91	82	73	64	57	49	41	33	24	17	9	1							
3	91	83	74	65	57	49	43	35	28	29	14	7							
4	92	83	75	67	59	51	43	35	32	25	18	11	4						
5	92	84	76	68	61	53	46	38	31	24	21	15	8	2					
6	92	85	77	70	62	55	48	41	34	27	20	14	12	6					
7	93	85	78	71	64	57	50	44	37	30	24	17	11	5					
8	93	86	79	72	65	59	52	46	39	33	27	21	15	9					
9	93	86	80	73	67	60	54	48	42	36	30	24	18	12	12	3			
10	93	87	81	74	68	62	55	50	44	38	33	27	21	16	15	6			
11	94	87	81	75	69	63	68	52	46	41	45	35	30	24	19				
12	94	88	82	76	70	65	59	54	48	43	37	32	27	22	21	12	4		
13	94	88	83	77	71	66	60	55	50	45	40	35	30	25					
14	94	89	83	78	72	67	62	57	52	47	42	37	32	27					
15	94	89	84	78	73	68	63	58	53	48	42	39	34	30	28	20	13	4	
16	95	89	84	79	74	69	64	59	55	50	43	41	37	32					
17	95	90	83	80	75	70	65	61	56	52	47	43	39	34					
18	95	90	85	80	76	71	66	62	57	53	49	45	40	36	35	27	20	13	6
19	95	90	86	89	76	72	67	63	59	54	50	46	42	38					
20	95	91	86	81	77	73	68	64	60	56	52	48	44	40	38	30	24		
21	95	91	86	82	78	73	69	65	61	57	53	49	45	42	39	32	26	19	13
22	95	81	87	82	78	74	70	66	62	58	54	50	47	43					
23	96	91	87	83	79	75	71	67	63	59	55	52	48	45					
24	96	91	87	83	79	75	71	68	64	60	57	53	49	46	43	37	31	26	21
25	96	92	88	84	80	76	72	68	65	61	58	54	51	47	44	38	33		
26	96	92	88	84	80	76	73	69	66	62	59	55	52	49					
27	96	92	88	84	81	77	73	70	66	63	59	56	53	50	47	41	36	31	26
28	96	92	88	84	81	77	73	71	63	65	61	58	55	53					
29	96	93	90	86	82	78	74	72	68	66	63	60	57	54					
30	96	93	90	86	82	79	76	73	70	67	64	61	58	55	50	44	39	35	30
33	96	93	90	86	83	80	77	74	71	68	66	63	60	57	52	47	42	37	33
35	96	93	90	87	83	80	77	74	71	70	66	64	60	59	54	50	44		
36	97	93	90	87	84	81	78	75	72	70		64		59	54	50	45	41	36
39	97	94	91	88	95	82	79	76	74	71		66		61	56	52	47	43	39
40		94		88						72		66		61	56	52	48		

- *Determination of Relative Humidity:*

In order to determine the Relative Humidity of a place, we have to record the temperature readings of the dry bulb and wet bulb. Then it is needed to calculate the difference of these two readings. Now from the table the relative humidity of a place can be determined.

Example 1

Temperature of dry bulb reading is 11°C
Temperature of wet bulb reading is 10.5°C
∴ Difference of the two readings = 0.5°C (Depression of the wet bulb)
∴ From the table it is found that the relative humidity of the place is 94% (Humidity is high)

Example 2

∵ Temperature of dry bulb reading is 11°C, and wet bulb reading is 4°C
∴ The difference of the reading is 7°C
∴ It is found from the table that the relative humidity of the place is 19% (Very dry).

Table 3: *Table for Determination of Relative Humidity*

Date: _______________ Place: _______________ Time: _______________

No. of Reading (N)	*Temperature of Dry Bulb Reading (°C) (a)*	*Temperature of Wet Bulb Reading (°C) (b)*	*Difference (a—b) is the Depression of the Wet Bulb*	*Mean Difference*	*Relative Humidity (RH) in % to be Determined on the Basis of Given Table*	*Remarks if Any*
1						
2						
3						

Records taken by:

Fortin's Barometer

A barometer means an instrument for measuring atmospheric pressure. Fortin's barometer is an improved barometer. It is light and portable. It consists of a vertical glass tube filled with mercury. It is closed in the upper end. The open end of this tube is inverted and immersed into a cistern of mercury. This cistern has a flexible bottom which can be adjusted by a screw. With the help of the screw, the zero level is adjusted. Above this level, the column of mercury is measured which indicates the air pressure. An ivory index is fixed at the zero level.

The glass-tube is fixed within a brass-tube on which a graduated scale is inscribed. The mercury level in the tube can be seen through a slit. There is a sliding Vernier (scale) on it for ensuring accurate reading. Vernier can easily be adjusted by screw. There is an attached thermometer on the barometer. It helps to make a temperature correction.

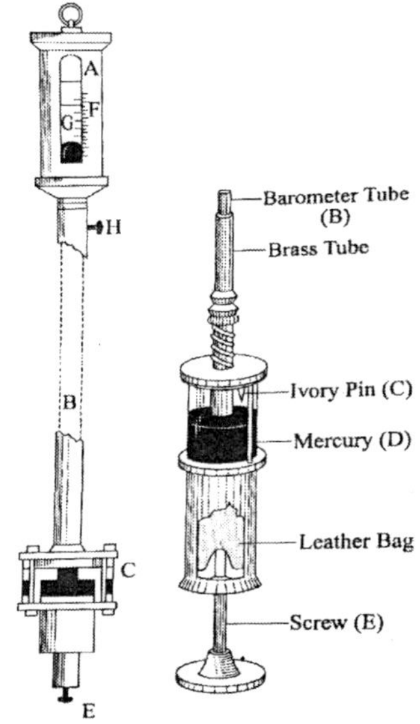

Figure 5: *Fortin's Barometer*

- *Reading of Barometer*

In order to have a reading of Barometer, it is needed to see whether the ivory pointer C just touches the mercury surface in the reservoir D or not. If it does not, then adjustment is needed. Then there is gap between the tip of the pointer and its image formed by the shining mercury surface. If, on the other hand, the tip dips into the mercury, a dimple will be formed there. Hence adjusting the mercury surface by the screw E so that the ivory pointer C just touches it as needed. This happens when the tip of the pointer just coincides with its image in the mercury. It is known as zero-adjustment, because by this adjustment the surface of mercury in the reservoir comes in level with zero-mark of the scale F.

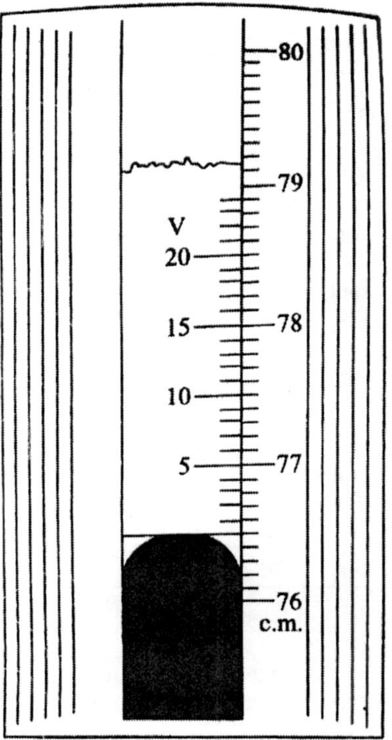

Figure 6: *A Portion of the Barometer showing the Vernier Scale and Reading*

Then, by turning the screw H, raise the Vernier G to a certain height above the level of mercury in the tube and then slowly lower it.

Now with the eye at the level of the convex surface of the mercury meniscus and slowly sliding down the Vernier till its lower edge is tangential to the convex mercury meniscus is needed. To do this adjustment correctly, a milk-white plate is attached on the wooden board behind the Vernier. As soon as the Vernier is adjusted correctly, the milk-white plate (being obstructed completely by the Vernier) will disappear from there. It is known as Vernier adjustment.

Then reading can be made. Read the Main scale and the Vernier scale avoiding parallax. Repeat the observations at least thrice; now the mean height of mercury column is found. The temperature from the thermometer attached with the barometer can also be noted.

- *Determination of Vernier Constant and Reading*

It is necessary to determine the Vernier Constant (VC); because it is different instrument. Suppose, if there are 10 small divisions in main scale, and if the 10 small divisions of Vernier scale are equal to 9 small divisions of main scale, then Vernier Constant would be:

$$1 \text{ divisions of Vernier Scale} = \frac{9}{10} \text{ divisions of main scale}$$

$$\therefore \text{ VC} = 1 - \frac{9}{10} \times 0.1 \text{ (small division of main scale)}$$

$$= \frac{10-9}{10} \times 0.1$$

$$= \frac{1}{10} \times 0.1$$

$$= \frac{1}{100} = 0.01\text{cm}$$

After determining the VC, proceeding to have the reading of main scale is needed. Suppose, the reading of the main scale is 75.6cm; but the height of the mercury is somewhat higher, then need to determine it from Vernier scale. Here in this case, the Vernier scale touches the main scale at 13. Hence the Vernier reading would be 13×0.01 = 0.13. It should now be added to the main scale reading. Therefore, the atmospheric pressure of this is 75.6 + 0.13 = 75.73cm or 757.3mm

To make accurate reading, taking reading thrice is needed and the average is considered. It is shown in the following table:

Table 4: *Calculation of Atmospheric Pressure*

Calculation Table To Measure The Atmospheric Pressure

Date: ________________ Place: ________________ Time: ________________

No. of Reading	*Main Scale Reading (a)*	*Vernier Constant*	*Vernier Reading*	*Vernier Reading × V.C (b)*	*Total Reading (a + b)*	*Mean Height (cm)*	*Mean Temperature*
1	75.6		13	0.01×13 = 0.13	75.6+0.13=75.73		
2	75.6	0.01	14	0.01×14 = 0.14	75.6+0.14=75.74	75.73	29°C
3	75.6		13	0.01×13 = 0.13	75.6+0.13=75.73		

Records taken by:

∴ At 29°C temperature, atmospheric pressure is:

$$\frac{75.73 + 75.74 + 75.73}{3}$$

$$= \frac{227.20}{3} = 75.73 \text{ cm or } 757.3 \text{ mm}$$

- *Method of Transferring mm (millimetre) to mb (miillibar)*

Given, 1000 dynes = 1 millibar

And 1000 millibars = 1 bar

1 bar = 29.53″ or 76 cm or 760 mm

∵ 760 mm = 1 bar

$$\therefore 1 \text{ mm} = \frac{1}{760} \text{ bar}$$

$$\therefore 757.3 \text{ mm} = \frac{1 \times 757.3}{760} = \frac{757.3}{760}$$

= 0.996 bar = 0.996× 1000 = 996 mb

This should be noted that the atmospheric pressure has an inverse relation with atmospheric temperature. The higher the temperature, the lower is the pressure. The general rule is that a rising thermometer shows a falling barometer, whereas a falling thermometer shows a rising barometer.

The atmospheric pressure is shown in the map by means of millibar (mb). It is shown in the weather map as isobars. Pressure gradients determine the direction and velocity of wind.

Rain Gauge

The rainfall of a place is measured by an instrument called rain- gauge. It consists of 2 parts, a collecting can and a measuring glass. The collecting can A is placed in an upright cylinder opened at the top containing a funnel BCD which drains into the can below. Any rain which falls over the mouth of cylinder runs down in the can is collected. At fixed intervals, the water is emptied out and the amount is measured in the graduated glass E. Though the rainfall is measured in inches, India has adopted a metric unit and the unit of measurement is mm. When we say, the mean annual rainfall of a place is 50 mm, we mean that if all the rain that falls upon that particular place during a normal year collected without the least amount being lost by evaporation or flowing away. The amount of water would cover the area to the depth of 50 mm. the line joining places having same mean monthly or yearly rainfall is called isohyets.

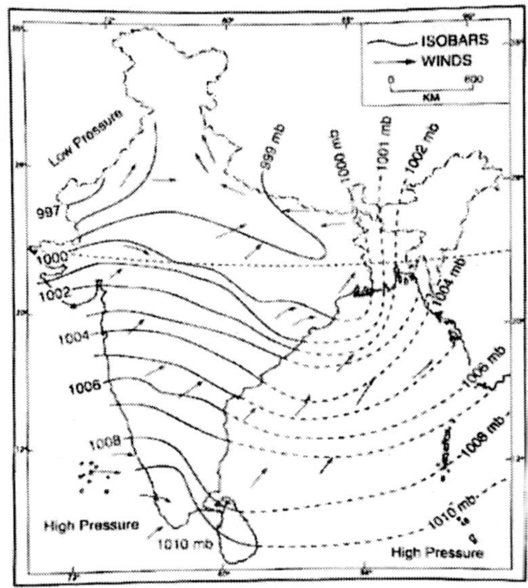

Figure 7: *Millibars are shown in the Weather Map of India*

Wind Vane

The instrument widely used for measuring wind direction is wind vane or weather cock. As wind direction is always blocked by trees and tall buildings wind-vanes or cocks need to be erected in an exposed position to get a five direction. It is made up of 2 parts – A

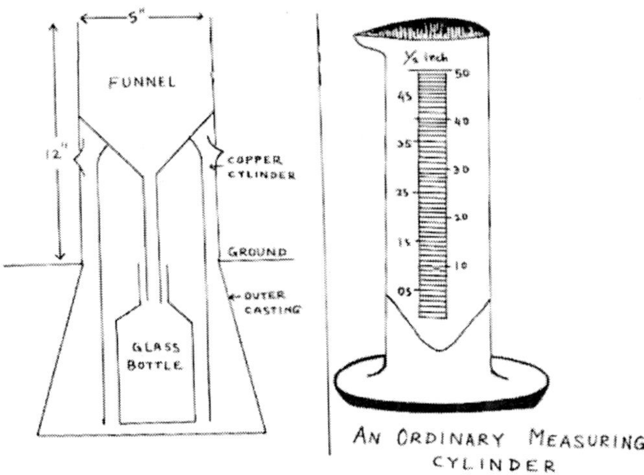

Figure 8: *Rain Gauge*

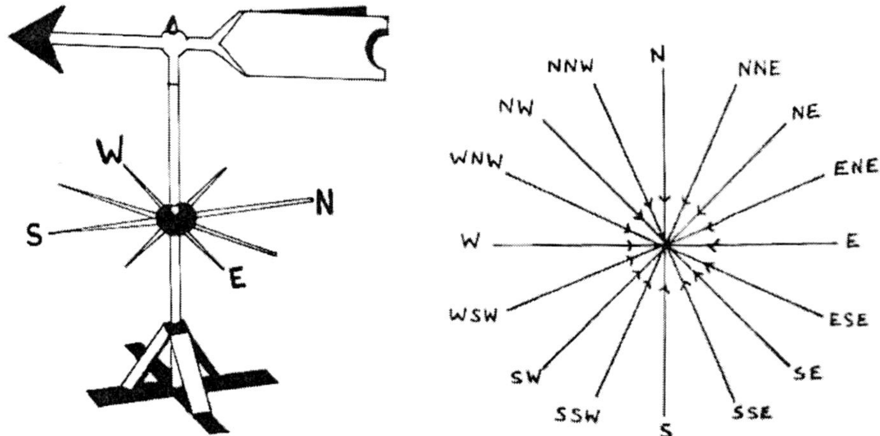

Figure 9: *Wind Vane and Wind Directions*

Table 5: *Table Showing Various Directions of Wind and Related Codes*

Code No.	*Description*	*Code No.*	*Description*
00	Calm	20	South–South West
02	North–North East	23	South West
05	North East	25	West–South West
07	East–North East	27	West
09	East	29	West–North West
11	East–South East	32	North West
14	South East	34	North–North West
16	South–South East	36	North
18	South	39	Variable

and B. A is one part and i.e., the arrow or vane on the top which is free to move with the prevailing wind. The other part (B) with the 4 compass points is stationary and shows in which direction is wind moving. Winds are always named from the direction they blow. A south-west wind is one that blows from the south-west.

6

Cartograms

Cartograms are the diagrammatic representation of numerical information and statistical data. The word 'carto' means diagram and 'gram' means making or composing; thus, it means 'diagram making'. According to the characteristics of the data it needs to be suitably represented by an appropriate graphical method and thereby meet the purpose of research and study. The authenticity of the data will lead to the accuracy of the graphical representation and interpretation.

Simple Bar Diagram

Simple Bar or Column diagrams consist of a series of columns or bars proportional in length in accordance to the represented quantities. If the width of the bars remains the same, they are called a one-dimensional diagram. When the data with only one component is available and each bar represents a single value, simple bar diagrams are drawn. A bar graph can help to identify the trend of data better than in a tabulated form.

Importance

- Comparison: Simple Bar Graph is quite an effective tool used for the purpose of convenient and quicker comparison among observations based on a statistical parameter.
- Convenience: It is easily understandable as information can be retrieved by observing the chart at a glance.
- Help in Decision Making: The bar graph can help to make a quick summary of a large set of data in a visual form, which may help to take any crucial decision.
- The trend of Data: A bar graph can help to identify the trend of data better than in a tabulated form.

Table 1: *The Calculation for Representing the Total Population of Different Districts of West Bengal, 2011*

Sl No.	*Districts of West Bengal*	*Total Population (m = in million)*	*Population Reduced to Scale (1 cm ≡ 3 million)*
1.	Darjeeling	1846823 = 1.84 m	0.61
2.	Jalpaiguri	3872846 = 3.87 m	1.29
3.	Cooch Behar	2819086 = 2.82 m	0.94
4.	Uttar Dinajpur	3007134 = 3.01 m	1.00
5.	Dakshin Dinajpur	1676276 = 1.68 m	0.56
6.	Malda	3988845 = 3.99 m	1.33
7.	Murshidabad	7103807 = 7.10 m	2.37
8.	Birbhum	3502404 = 3.50 m	1.17
9.	Burdwan	7717563 = 7.71 m	2.57
10.	Nadia	5167600 = 5.17 m	1.72
11.	North 24 Parganas	10009781 = 10.01 m	3.34
12.	Hooghly	5519145 = 5.52 m	1.84

Sl No.	*Districts of West Bengal*	*Total Population (m = in million)*	*Population Reduced to Scale (1 cm ≡ 3 million)*
13.	Bankura	3596674 = 3.60 m	1.20
14.	Purulia	2930115 = 2.93 m	0.98
15.	Howrah	4850029 = 4.85 m	1.62
16.	Kolkata	4496694 = 4.50 m	1.50
17.	South 24 Parganas	8161961 = 8.16 m	2.72
18.	Paschim Medinipur	5913457 = 5.91 m	1.97
19.	Purba Medinipur	5095875 = 5.09 m	1.70

Source: State Statistical Handbook, 2014, Bureau of Applied Economics and Statistics, Government of West Bengal

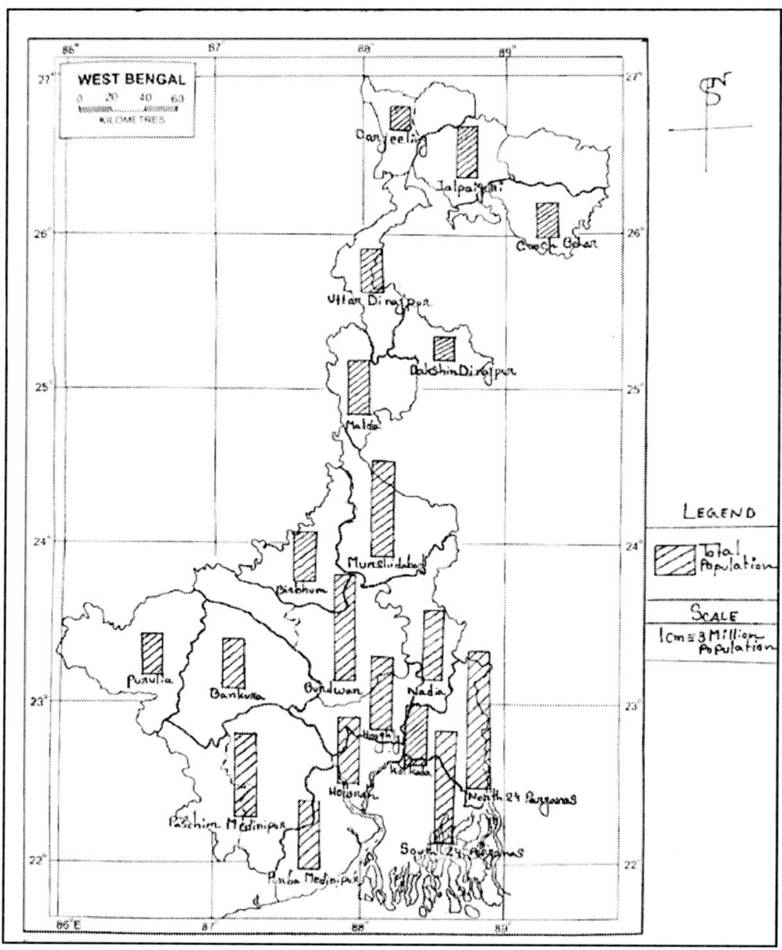

Figure 1: *Simple Bar Diagram showing the Total Population of Different Districts of West Bengal (2011)*

Source: Developed by the author.

Interpretation

According to the given data on the population of West Bengal (2011), it is seen that the simple bar graphs represent the distribution of the district-wise population of West Bengal.

From the map, it is seen that the highest populated district is North 24 Parganas (10.01 million). Bankura, Howrah, Hooghly, Purulia, Kolkata, Nadia, and Murshidabad consist of a medium range of population (4 to 6 million population). The least populated districts are Darjeeling, Cooch Behar, Malda, Jalpaiguri, Birbhum, etc. Dakshin Dinajpur has the lowest share of the population (1.68 million) among all the districts of West Bengal.

In short, it can be said that the total population of the northern districts is much lower than the southern districts of West Bengal. The northern districts consist of rocky and hilly terrain, infertile soil, and forested zones and this has resulted in lesser-developed infrastructure and other civic facilities in comparison to other regions. Thereby, these districts represent a low population. Whereas the southern districts possess fertile low-lying agricultural tracts, varied industrial plants, favourable transport and communication network, well-developed infrastructure, etc. This has resulted in the development of the tertiary sector of the economy and the generation of employment which has ultimately accentuated the growth of the population in these regions.

Multiple Bar Diagram

In a *multiple bar diagram or multiple bar graph,* two or more sets of interrelated data are represented using two sets of horizontal bars. Multiple bar diagram facilitates comparing or exhibiting the relationship between two or more two sets of data. The multiple bar graphs are drawn similar to simple bar graphs but in this case, for two bars two different shades, colors, or dots are used to represent two distinct sets of data.

The basic principle of drawing a multiple bar diagram is that the length of the bars is proportional to the value of the representable item. Depending on the nature of the data, multiple bar diagrams are drawn when different bars of proportionate lengths are drawn side by side to represent the constituents. Multiple bar graphs are also one-dimensional diagrams.

Importance

- Comparison: It can be applied to contrast two or more two sets of data.
- Establishing Relationship: Through the multiple bar diagram relationship between two or more phenomena inherited can be established.
- Information Retrieve: A single diagram provides information about two or more variables just side by side.
- Simplicity: It looks visually appealing and is very simple to understand.

Table 2: *The Calculation for Representing Male-Female Population of Different Districts of West Bengal (2011)*

		Total Population		*Population Reduced to scale (1 cm = 20,00000)*	
Sl No.	*Districts of West Bengal*	*Male*	*Female*	*Male*	*Female*
1.	Darjeeling	937259	909564	0.47	0.45
2.	Jalpaiguri	1983064	1889782	0.99	0.94
3.	Cooch Behar	1451542	1367544	0.73	0.68
4.	Uttar Dinajpur	1551066	1456068	0.78	0.73

		Total Population		Population Reduced to scale (1 cm = 20,00000)	
Sl No.	*Districts of West Bengal*	*Male*	*Female*	*Male*	*Female*
5.	Dakshin Dinajpur	857199	819077	0.43	0.41
6.	Malda	2051541	1937304	1.03	0.97
7.	Murshidabad	3627564	3476243	1.81	1.74
8.	Birbhum	1790920	1711484	0.90	0.86
9.	Burdwan	3966889	3750674	1.98	1.88
10.	Nadia	2653768	2513832	1.33	1.26
11.	North 24 Parganas	5119389	4890392	2.56	2.45
12.	Hooghly	2814653	2704492	1.41	1.35
13.	Bankura	1838095	1758579	0.92	0.88
14.	Purulia	1496996	1433119	0.75	0.72
15.	Howrah	2500819	2349210	1.25	1.17
16.	Kolkata	2356766	2139928	1.18	1.07
17.	South 24 Parganas	4173778	3988183	2.09	1.99
18.	Paschim Medinipur	3007885	2905572	1.50	1.45
19.	Purba Medinipur	2629834	2466041	1.31	1.23

Source: State Statistical Handbook, 2014, Bureau of Applied Economics and Statistics, Government of West Bengal

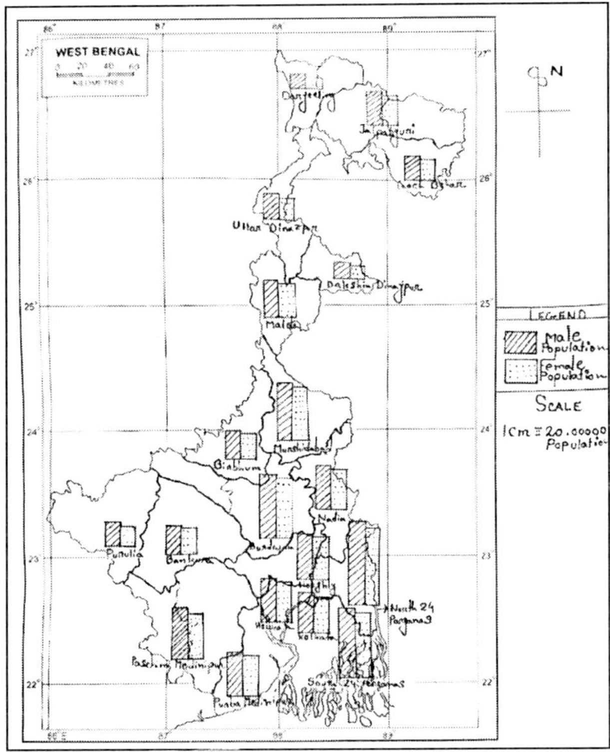

Figure 2: *Multiple Bar Diagram showing the Male-Female Population of Different Districts of West Bengal (2011)*

Source: Developed by Author

Interpretation

The multiple bar diagrams have been drawn with the district-wise male and female populations of West Bengal. It can be observed that the trend of the male population is higher in all of the districts. The male-female difference is more in the districts of Burdwan, Howrah, Kolkata, Bankura, North 24 Parganas, etc. The male population had moved to these urban sites in search of jobs which has led to the stark male-female difference as their families remain in the rural native land. Comparatively, the difference is low in Hooghly, Paschim Medinipur, Birbhum, and Murshidabad where the availability of agricultural land, agro-based industries, and other small scale-industries generate employment for the population. Again, the mountainous, and Uttar and Dakshin Dinajpur districts exhibit a parity between male and female populations.

Compound Bar Diagram

Compound Bar Diagrams are drawn when the data comprises more than one component within a total and each bar is proportionately divided to show the constituents as well as the total. Each category is distinguished by different shades. A suitable scale for all the bars should be selected for the drawing and appropriate presenting the components

Importance

- Division in totality: This bar diagram is used to represent data in which the total magnitude is divided into different components.
- Retrieve information: Compound Bar Diagram helps to retrieve information not only one component rather its sub-parts at a glance.
- Visual impression: It is very easy to understand for any layperson as the number of elements is proportionate to the height of the bar and within each bar, the sub-parts are shown in different shades/colours or hachures.
- Easy comparison: A comparison of elements existing in different areas can be done using this diagram.
- Summarise large data set: It helps to summarise huge data sets with multiple sub-parts and reaching to any conclusion.

Table 3: *The Calculation for Representing Rural-Urban Population of Different Districts of West Bengal (2011)*

		Total Population			*Population Reduced to Scale (1cm = 3 Million)*		
Sl No.	*Districts of West Bengal*	*Rural*	*Urban*	*Total*	*Rural*	*Urban*	*Total*
1	Darjeeling	1118860	727963	1846823	0.37	0.24	0.62
2	Jalpaiguri	2812495	1060351	3872846	0.94	0.35	1.29
3	Cooch Behar	2529652	289434	2819086	0.84	0.10	0.94
4	Uttar Dinajpur	2644906	36228	2681134	0.88	0.01	0.89
5	Dakshin Dinajpur	1439981	236295	1676276	0.48	0.08	0.56
6	Malda	3447185	541660	3988845	1.15	0.18	1.33
7	Murshidabad	5703115	1400692	7103807	1.90	0.47	2.37
8	Birbhum	3052956	449448	3502404	1.02	0.15	1.17
9	Burdwan	4639264	3078299	7717563	1.55	1.03	2.57
10	Nadia	3728727	1438873	5167600	1.24	0.48	1.72
11	North 24 Parganas	4277619	5732162	10009781	1.43	1.91	3.34
12	Hooghly	3390646	2128499	5519145	1.13	0.71	1.84
13	Bankura	3296901	299773	3596674	1.10	0.10	1.20

Sl No.	Districts of West Bengal	Total Population			Population Reduced to Scale (1cm = 3 Million)		
		Rural	Urban	Total	Rural	Urban	Total
14	Purulia	2556801	373314	2930115	0.85	0.12	0.98
15	Howrah	1775885	3074144	4850029	0.59	1.02	1.62
16	Kolkata	0	4496694	4496694	0.00	1.50	1.50
17	South 24 Parganas	6074188	2087773	8161961	2.02	0.70	2.72
18	Paschim Medinipur	5190771	722686	5913457	1.73	0.24	1.97
19	Purba Medinipur	4503161	592714	5095875	1.50	0.20	1.70

Source: State Statistical Handbook, 2014, Bureau of Applied Economics and Statistics, Government of West Bengal

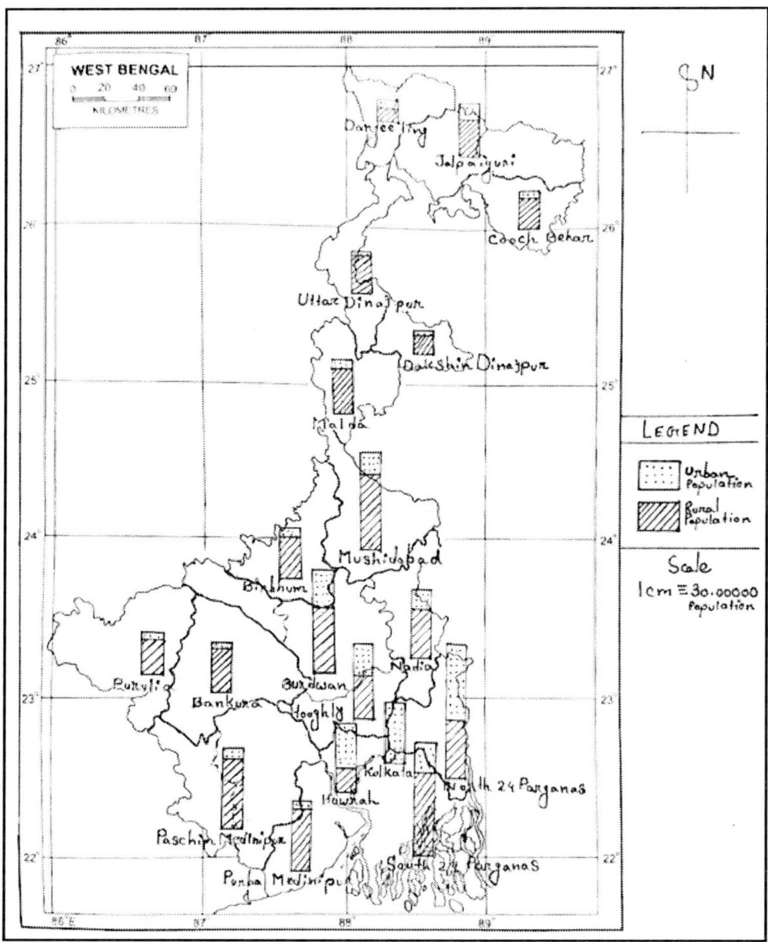

Figure 3: *Compound Bar Diagram showing the Rural and Urban Populations of Different Districts of West Bengal (2011)*

Source: Developed by Author

Interpretation

The compound bar graphs have been drawn with the district-wise rural and urban populations of West Bengal. The trend of the rural population is high in most of the districts of West Bengal.

Kolkata records no rural population. Infact this megacity is the seventh-most populous city in India(2011). Other districts like North 24 Parganas, Howrah exhibit high urban population in comparison to the rural total. The southern districts possess a greater share of the urban population and bear the facilities of diversified occupational structure, infrastructure, transport, communication and other services.

In the northern districts, especially Darjeeling, Uttar Dinajpur, Dakshin Dinajpur, etc. rural population is much higher than the urban population. Most of the people are engaged in agricultural activities, possess lesser diversity in occupational opportunities and the civic facilities are in the stage of developing.

Pie Diagram

Pie diagrams are represented by proportional circles which are further divided into sectors to present associated items into these sectoral parts. To calculate the radii of the proportional circles the area of the total individual value is taken to represent the circle i.e. πr^2. Thus the radius can be obtained by dividing the total value by π and then finding the square root. Sometimes to simplify the calculation only the square roots are calculated since π is common. The radii are then reduced to scale and drawn for suitable comparison.

To divide each circle into its sectors, the total value is taken to represent 360° and the various sectors are calculated in their angular degrees to the total. To give a visual effect the sectors are then distinctly coloured or shaded.

Importance

- Easy to understand: It is a simple picturesque representation that can be easily understood.
- Comparison: Through this diagram not only proportionality of any element can be represented but also its inherited sub-division or sector wise analysis can be derived.
- Decision Making: A pie diagram helps to take any critical decision as a long set of data with diverse sub-divisions/sectors can be represented in visual form.

Interpretation

Pie graphs have been drawn to show the type of total workers engaged in economic activities of six selected districts of West Bengal (2011). In terms of total workers, South 24 Parganas has the highest number of total workers and Purulia has the lowest. In this data series, the workers are subdivided into 3 broad divisions, those are: i) Agricultural workers, ii) Non-agricultural workers, iii) Handloom or handicraft or footloose industry workers. After representing the pie diagram in respective districts, it is seen that the share of agricultural workers is much lower than non-agricultural-sector workers in all the districts of West Bengal. It can be observed that the share of the workers at the non-agricultural sector is strikingly highest in all the six districts as urbanisation is gradually taking place. The agricultural-sector shows the least share in comparison to the Handloom/ handicraft or footloose industry with the exception of Jalpaiguri, Paschim Medinipur, and Purulia. These three districts possess agricultural land which results in considerable workers engaged in this sector. As Jalpaiguri is in the hilly terrain, the handloom/ handicraft/ Footloose industry has a very low or meagre percentage of workers. Whereas, with the changing land-use pattern, all the districts

Table 4: *The Calculation for representing the Type of Total Workers engaged in Economic Activities of Six selected Districts of West Bengal (2011)*

		Type of Workers						*Area in Degree*		
Sl No.	*Districts*	*Agricultural Sector*	*Non-Agricultural Sector*	*Handloom/ Handicraft/ Foot Loose Industry*	*Total Workers*	*Pie Radii*	*Radius in Scale (1 cm = 400 units)*	*Agricultural Sector*	*Non-Agricultural Sector*	*Handloom/ Handicraft/ Foot Loose Industry*
1	Jalpaiguri	46249	265717	4415	316381	317.34	0.79	52°37′	302°22′	5°01′
2	Nadia	39426	300877	51698	392001	353.24	0.88	36°13′	276°19′	48°28′
3	Howrah	15577	309397	52854	377828	346.79	0.87	14°50′	294°48′	50°22′
4	South 24 Parganas	29395	468686	41487	539568	414.43	1.03	19°37′	312°42′	27°41′
5	Paschim Medinipore	76212	261754	14713	352679	335.05	0.84	77°48′	267°11′	15°01′
6	Purulia	21239	108835	7135	137209	208.99	0.52	55°44′	285°33′	18°43′

Source: Economic Census of West Bengal, Government of India, 2012-13.

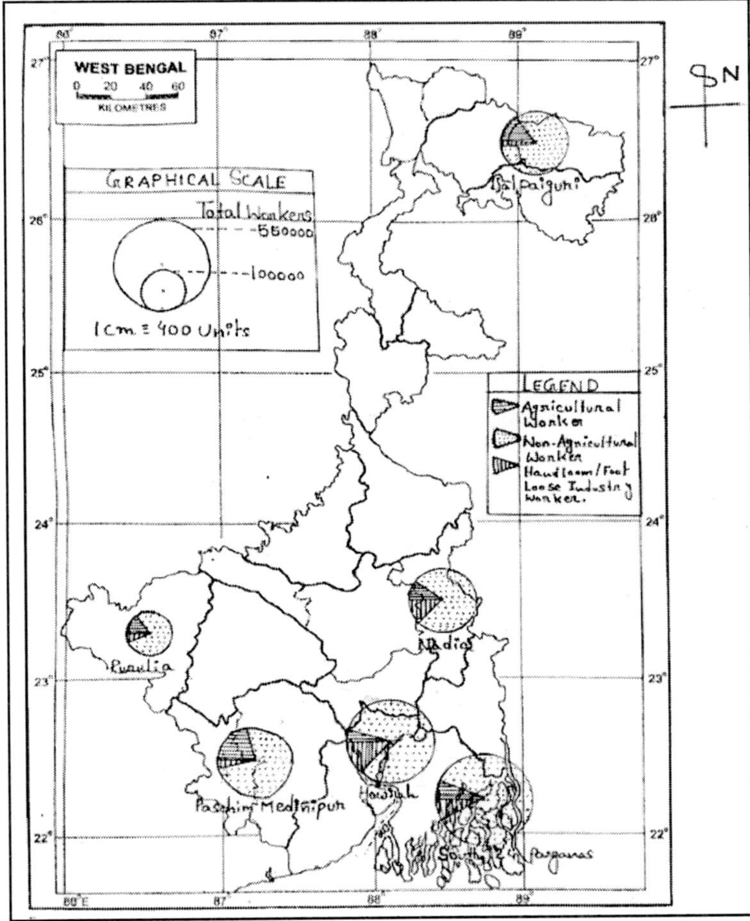

Figure 4: *Pie graphs showing the Type of Total Workers engaged in Economic Activities of Six selected Districts of West Bengal (2011)*

Source: Developed by author

draw huge workers at non-agricultural sector which encompasses trade and commerce, business, transport and communication, manufacturing, construction, and other services. In the future, the scenario of the workers engaged in the Handloom/ handicraft or footloose industry might change with governmental intervention.

Choropleth

Choropleth is a statistical thematic map that gives easy visualization of the range of intensity of the variable or element. The word "Choropleth" has been derived from two Greek words – "Choros" meaning 'area/ region' and "Plethos" meaning 'multitude'. Thus, the literal meaning of choropleth is "quantity in the area". Technically, choropleth is a quantitative areal map that shows the spatial distribution of intensity or density of any element with the help of a system of graded shading or colour drawn within the boundaries of the administrative units.

According to Monk House and Wilkinson (1826), "if the isopleth is the chief tool to a climatologist, the choropleth may be said to be the chief tool for human geographers." This method is primarily used to show the density of commodities per unit area, the average measure of percentages, etc., for example, the density of population, male-female ratio (sex ratio), percentage of agricultural lands under different crops to total land, or percentage of irrigated (types used) land to cultivated land or other physical and cultural geographical elements.

Importance

- Spatial distribution of elements: Through choropleth mapping, the spatial distribution of any element can be represented effectively. The intensity of shading is directly proportional to the density of the element.
- Easy to understand: These maps are developed using shading variation proportional to the distribution of any element so that it is easily understandable and one can retrieve information from the map.
- Comparative distribution: This is a simple method that gives comparative information about the distribution of an element over two or more places at a glance.
- To represent an abrupt change in element: Shading method is especially suited to those elements in which there is an abrupt change. For example – the density of the population.
- Resolve misconception: The misconception that commonly arises in the drawing of an isopleth map is the grouping of a unit under one average value, which can be resolved through the choropleth technique as the density zones are based on administrative units, reveals the average distribution of the element of the particular administrative region.

Table 5: *The Calculation of Choropleth for representing the Density of Population in Different Districts of West Bengal, 2011*

Sl. No.	*Name of Districts*	*Area in sq. km.*	*Total Population 2011*	*Density/sq.km. (Calculated)*	*Density/sq.km. (Actual)*
1.	Darjeeling	3149	1846823	586.479	586
2.	Jalpaiguri	6227	3872846	621.944	622
3.	Cooch Behar	3387	2819086	832.325	832
4.	Uttar Dinajpur	3140	3007134	957.686	958
5.	Dakshin Dinajpur	2219	1676276	755.420	755
6.	Malda	3733	3988845	1068.536	1069
7.	Murshidabad	5324	7103807	1334.299	1334
8.	Birbhum	4545	3502404	770.606	771
9.	Burdwan	7024	7717563	1098.742	1099
10.	Nadia	3927	5167600	1315.915	1316
11.	North 24 Parganas	4094	10009781	2444.988	2445
12.	Hooghly	3149	5519145	1752.666	1753
13.	Bankura	6882	3596674	522.620	523
14.	Purulia	6259	2930115	468.144	468
15.	Howrah	1467	4850029	3306.087	3306
16.	Kolkata	185	4496694	24306.454	24306
17.	South 24 Parganas	9960	8161961	819.474	819
18.	Paschim Medinipur	9368	5913457	631.240	631
19.	Purba Medinipur	4713	5095875	1081.238	1081

Source: State Statistical Handbook, 2014, Bureau of Applied Economics and Statistics, Government of West Bengal

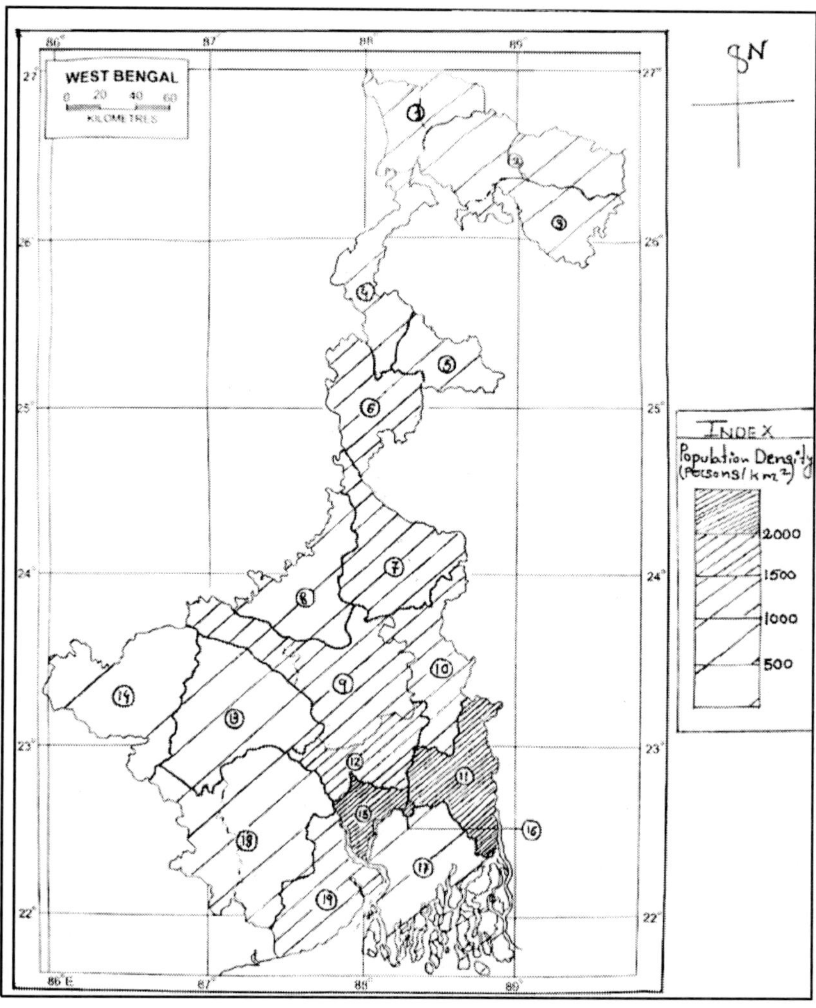

Figure 5: *Choropleth Map showing Population Density of Different Districts of West Bengal*

Source: Developed by the author

Calculation for the Division of Classes/Class Intervals with Population
Highest density = 24306.45
Lowest density = 468.14
Class Range (Highest-Lowest) = (24306.45-468.14) = 23838.31
Number of classes = 5
Class Interval will be = 23838.31 / 5 = 4767.662

For a more or less, equal representation of the districts in all the Class Intervals, it is judicious to take Class Intervals at 500 instead of 5000.

∴ Class Intervals taken are: i) <500, ii) 500-1000, iii) 1000-1500, iv) 1500-2000, v) >2000

Table 6: *Name of Districts Based on Serial No. Placed on Map*

Darjeeling (1)	Jalpaiguri (2)	Cooch Behar (3)	Uttar Dinajpur (4)	Dakshin Dinajpur (5)	Malda (6)	Murshidabad (7)
Birbhum (8)	Burdwan (9)	Nadia (10)	North 24 Parganas (11)	Hooghly (12)	Bankura (13)	Purulia (14)
Howrah (15)	Kolkata (16)	South 24 Parganas (17)	Paschim Medinipur (18)	Purba Medinipur (19)		

Table 7: *Name of Districts under Different Population Density Classes*

Class Division (Based on Population Density)	*Density Zones*	*Name of Districts*
>2000	High	Kolkata, Howrah, North 24 Parganas
1500-2000	Medium	Hooghly
1000-1500		Malda, Murshidabad, Burdwan, Nadia, Purba Medinipur
500-1000	Low	Darjeeling, Jalpaiguri, Cooch Behar, North Dinajpur, South Dinajpur, Birbhum, Bankura, Paschim Medinipur, South 24 Parganas
<500		Purulia

Source: Developed by Author Based on Population Density Map of West Bengal, 2011

Interpretation

Population density is a measurement of population per unit of land area. The density of the population varies based on the geographical location, the scope of occupation, civic services for better living, amenities of education, available infrastructure, urbanization trend and stability of the political regime, etc.

The 5 distinct zones under the Population Density mapping of West Bengal can be broadly classified under 3 distinct density zones, namely Low (Below 1000), Medium (Between 1000 to 2000), and High (Above 2000) Density Zone.

High population density is noticed in 3 districts of West Bengal, namely Kolkata, Howrah, and North 24 Parganas. Among them the highest and distinct density has been recorded in Kolkata, more than 24000 persons per sq. km. Kolkata, the capital of West Bengal by the side of River Ganga is one of the populous metropolitan areas possessing all the urban civic facilities to its maximum. This is the main hub with business, education, administration, and medical facilities. Howrah and North 24 Parganas are two adjacent districts of Kolkata which are also very developed and hold a high concentration of population.

Relatively medium density of population has been observed in Hooghly, Malda, Nadia, Purba Medinipur, Burdwan and Murshidabad. With the availability of fertile agricultural land and agro and ancillary-based industries, a comparatively appreciable amount of the population is found in these regions. The primary sector provides huge employment and to some extent the secondary and the developing tertiary sectors.

The districts like Purulia, Bankura, Darjeeling, North Dinajpur, South Dinajpur have low population primarily due to the harsh physical location, climate, and infertile soil. South 24 Parganas, Paschim Medinipur, Jalpaiguri, Cooch Behar, and Birbhum are developing amidst unfavourable geographical conditions. The basic transport and communication system, civic amenities, and infrastructural outlay are yet to be developed to their maximum which would then surely draw a greater population.

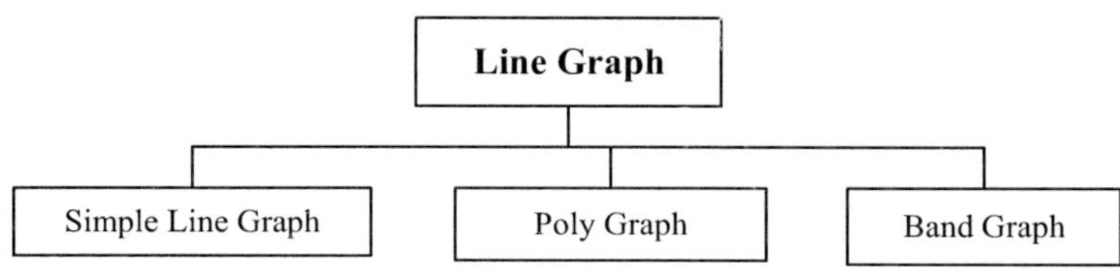

Figure 6: *Different Types of Line Graph*

With the help of choropleth mapping in different districts of West Bengal, the difference in regional developmental conditions truly reflects on the density distribution.

Line Graph

A *line graph* may be defined as a graph represented by a line or a series of lines. It precisely represents some statistical data and information. There are different types of line graphs. These may be grouped into the following three types:

Simple Line Graph

In a simple line graph, a single graphical line is drawn according to the value put forwarded by statistical data.

Methods of Drawing Line Graph

1. Drawing a line on the bottom of a graph (at abscissa) and dividing it into equal parts as many as one requires to show the divisions of days/months/years/or other units.
2. On the two definite ends of the line (abscissa), erecting two vertical lines (on the ordinates) as shown in the figure and dividine them to show the scale (of temperature, rainfall, amount of production, population growth, or other items given on the statistical table).
3. Now plotting or marking the determining coordinates along X-axis and Y-axis on the graph by points.
4. Then joining the points by a straight line (curve line) to show the line graph (curve).

Table 8: *The Calculation of Simple Line Graph for representing the Urban Population of India (1901- 2011)*

Year	*Urban Population in Million*	*Population Reduced to Scale 1 cm ≡ 5 Millions*
1901	2.5	0.5
1911	2.6	0.52
1921	2.8	0.56
1931	3.3	0.66
1941	4.4	0.88
1951	6.2	1.24
1961	7.9	1.58
1971	10.9	2.18
1981	15.9	3.18
1991	21.4	4.28
2001	28.5	5.7
2011	37.7	7.54

Source: Census of India, 2011

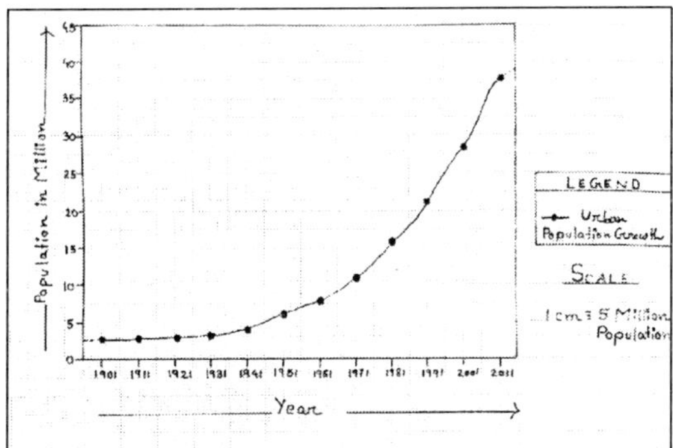

Figure 7: *Simple Line Graph showing Urban Population Growth of India (1901-2011)*

Source: Developed by Author

Interpretation: The simple line graph represents the statistical data in a single graphical line. After drawing the line graph showing the growth trend of urban population of India from 1901 to 2011 it is studied that in the pre-independence period it was very negligible, but in post-independence especially, after 1991 the growth of the urban population escalated tremendously. Due to the availability of urban civic facilities for education, services, infrastructure, administration, business and commerce, and medicine, the urban population has increased in India. The standard of living has changed, job opportunities extended, and a good communication network system has developed. With the growth of numerous urban centers, several challenges have also cropped up. With globalization, Indian urbanization would change immensely in the twenty-first century.

Polygraph

In a *polygraph*, there are a series of line graphs. They are known as multiple-line graphs. In this, more than one series of data can be represented in a single graph. Through this graph statistical data can not only be represented simultaneously but also, they can be compared.

Methods of Drawing Polygraph

1. The method of drawing a polygraph is the same as drawing of the line graph.
2. In the poly graph two or more than two variables are presented through more than one line graph.

Interpretation: The polygraph is such a graph that can place more than one series of data within a single graph. In the present study, the polygraph is showing the rural and urban population growth in India from 1901 to 2011 with two sets of line graphs separately. After plotting two data sets it is studied that the growth rate of the rural population is much higher than the urban population. The rural population is rising positively but its trend is negative (in 1991-2001 growth of the rural population is 11.3 million whereas in 2001-2011 growth of the rural population is 9.1 million). In contrast, the urban population is rising positively and its trend is also positive (in 1991-2011 growth of the urban population is 7.1 million whereas in 2001-2011 growth of the urban population increased to 9.2 million). In India, the trend of rural population growth exhibits a diminishing trend in comparison to the urban population.

Table 9: *The Calculation of Polygraph for representing the Rural-Urban Population of India (1901- 2011)*

Year	Population in Million		Population Reduced to Scale (1 cm ≡ 10 Million)	
	Rural	Urban	Rural	Urban
1901	21.3	2.5	2.13	0.25
1911	21.6	2.6	2.16	0.26
1921	22.3	2.8	2.23	0.28
1931	24.6	3.3	2.46	0.33
1941	27.5	4.4	2.75	0.44
1951	29.9	6.2	2.99	0.62
1961	36	7.9	3.6	0.79
1971	43.9	10.9	4.39	1.09
1981	52.4	15.9	5.24	1.59
1991	62.9	21.4	6.29	2.14
2001	74.2	28.5	7.42	2.85
2011	83.3	37.7	8.33	3.77

Source: Census of India, 2011

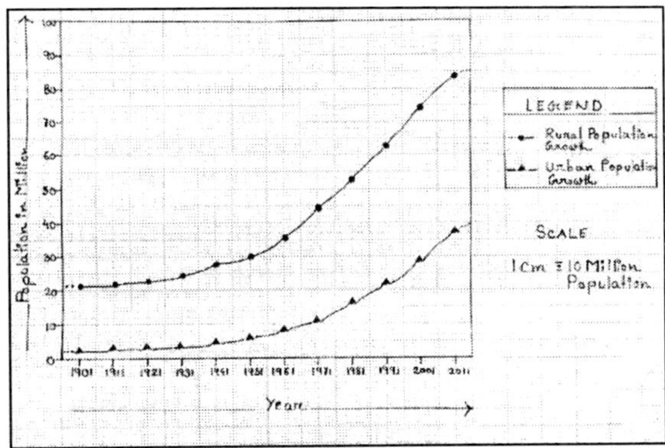

Figure 8: *Polygraph showing the Rural-Urban Population Growth in India (1901-2011)*

Source: Developed by Author

Band Graph

Band graph is also known as a compound line graph. In this graph, in between a series of line graphs, different shades are made between two-line graphs so that separations are made clearly. These may show the range and high-low value series of the variables studied.

Methods of Drawing Band Graph

1. The same method of drawing line graphs and polygraph are followed.
2. Then marking different shades in between the regions of line graphs so that differences can be easily made and layers can distinctively be shown.

Interpretation: A band graph is similar to a polygraph but the only difference is that in between line graphs separate shades are given to denote the area of separation. Herein, it is tried to represent the growth of the population of India from 1901 to 2011 in three distinct aspects i.e. total population growth, rural population growth, and urban

Table 10: *The Calculation of Band Graph for representing the Rural-Urban Population of India (1901-2011)*

	Population in Million			Population Reduced to Scale (1 cm ≡ 10 Million)		
Year	*Urban*	*Rural*	*Total*	*Urban*	*Rural*	*Total*
1901	2.5	21.3	23.8	0.25	2.13	2.38
1911	2.6	21.6	24.2	0.26	2.16	2.42
1921	2.8	22.3	25.1	0.28	2.23	2.51
1931	3.3	24.6	27.9	0.33	2.46	2.79
1941	4.4	27.5	31.9	0.44	2.75	3.19
1951	6.2	29.9	36.1	0.62	2.99	3.61
1961	7.9	36	43.9	0.79	3.6	4.39
1971	10.9	43.9	54.8	1.09	4.39	5.48
1981	15.9	52.4	68.3	1.59	5.24	6.83
1991	21.4	62.9	84.3	2.14	6.29	8.43
2001	28.5	74.2	102.7	2.85	7.42	10.27
2011	37.7	83.3	121	3.77	8.33	12.1

Source: Census of India, 2011

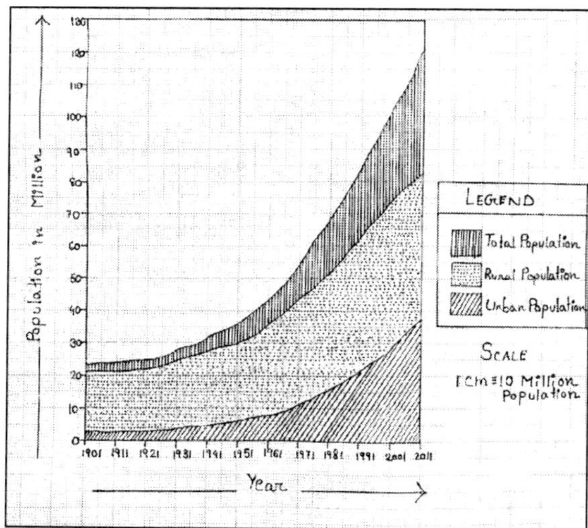

Figure 9: *Band Graph showing the Trend of Population Growth in India (1901-2011)*

Source: Developed by Author

population growth. After plotting all the sets of data, and giving shades it is seen that all three graphs exhibit a rising trend in India. Rural-urban migration has occurred for employment and a better standard of living opportunities. The presence of transport, communication, infrastructure, and services is leading to urban sprawl. With time these develop into cities and townships and attract a greater population and these regions tender all-inclusive facilities for the community.

Flow Maps/Charts

The Flow chart is a combination of graph and map. It is drawn to show the flow of commodities or people between the places of origin and destination. It is also called a Dynamic Map. Passenger and vehicle movement is represented through flow charts. In flow diagrams/charts the frequency of traffic flow between two areas is shown by the breadth of the line joining the two points (areas) on the map. The thickness of the line, the more will be the amount of traffic/passenger flow implying a developing region and *vice versa.* Generally, metropolia, industrial regions, busy cities, and centres of business and commerce show flow diagrams with thick lines. Sometimes flow diagrams are shown by parallel lines. In this case, the frequency of traffic flow is shown by the number of lines. The more the amount of flow of passengers or more will be the number of parallel lines and *vice versa.*

Example: Four stations A, B, C, and D are shown the flow of vehicles are shown between these stations. The daily number of vehicles is as follows:

Road	*Length in km.*	*Length in cm.* *1 cm = 8 km*	*Direction*	*Number of*	*Vehicles in Scale* *1 cm = 50 vehicles*
A-B	65	8.13	East	126	2.4
B-C	65	8.13	South-East	36	0.72
B-D	77	9.63	North-East	48	0.96

Interpretation

The flow diagram shows the maximum number of vehicles from A toward B which is 126. But the minimum number of vehicles moving from B to C is 36. The number of vehicle and passenger movement directly represent the level of development or the availability of facilities. Several planning agencies (government and NGOs) prepare flow charts to show the problems and prospects of that region. Consequently, developmental measures can be taken up and the regional disparities can be studied.

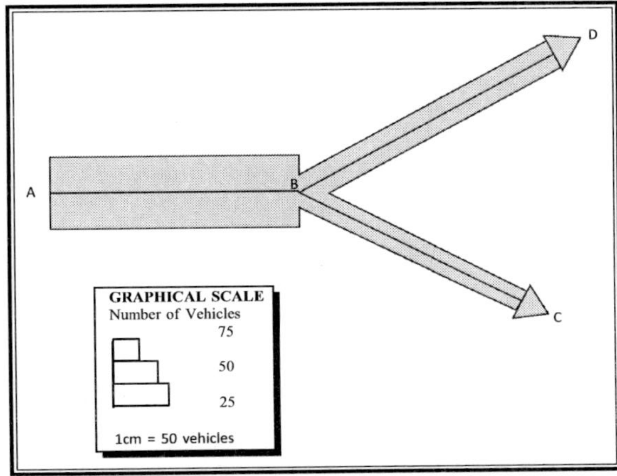

Figure 10: *Flow Maps/Charts showing Movement of Vehicles*

Star Diagram

A star diagram is used for organizing the characteristics of a single topic. A central space is used for displaying the topic, with each 'point' of the star listing some fact, attribute, or trait about the topic.

A Star diagram is also known as a clock diagram or vector diagram or direction diagram. Wind Rose is an example of a Star diagram. The wind rose shows the direction and intensity of wind for any particular region. From a point, straight lines are drawn in various directions. According to a given number of days, corresponding to each direction expressed in proper scale. Then these points are joined. A number of calm days, if any, are written in a circle at the centre.

Example: Below is the given temperature and rainfall of the months in a year to draw a Star diagram.

Month	*Jan.*	*Feb.*	*Mar.*	*April*	*May*	*June*	*July*	*Aug.*	*Sept.*	*Oct.*	*Nov.*	*Dec.*
Temperature in 0°C	28.9	27.8	25	20	15.6	12.2	11.7	14.4	18.3	22.8	26.1	27.8
Rainfall in cms.	4.2	3.3	2.8	1.0	1.5	1.2	0.8	0.8	0.8	1.7	3.0	3.8

Interpretation

On the basis of the above data, a Star diagram is drawn. From the diagram, it has been observed that the temperature is high during the month of January, February, March, November, and December, and the rainfall has also occurred about 4.2 cm in the month of January. But the rainfall and temperature are less in the month of June. So, it can be said that the given data is recorded from a region situated in the southern hemisphere as in this hemisphere the temperature is high during the month of December.

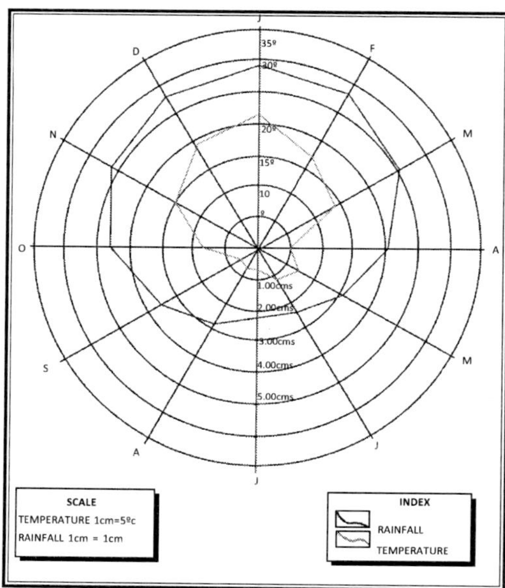

Figure 11: *Star Diagram showing Temperature and Rainfall*

7

Chorochromatic Map

The term "Chorochromatic" can be applied to a special category of the thematic map, where the quantity and values of the statistical data are not considered. This category of map is often called a "Colour patch map"(Monkhouse and Wilkinson, 1971), as the term colour is used in its widest interpretation of tints or shades. The purpose of this chorochromatic map is to identify and differentiate the different regions of a map based on colours. It can be commonly used in geography studies and also for many other studies related to land- use. Within the boundaries of the respective administrative unit, spatial patterns of the ground surface are outlined and then displayed with colour symbols such as tints or colours.

Importance

- *Land-use pattern-related information:* The chorochromatic maps are also known as land-use maps as they represent the quantity of land utilization and distribution of different natures of land-use and other types of commodities. These maps can be used for regional planning,
- *Use of conventional colours, shades, and signs*: Functional chorochromatic maps or morphology of urban centres are prepared through the selection of some conventional colours or signs. These may be yellow for cultivation lands, green for forest, brown for fallow, red for settlements, metalled roads, un-metalled roads, cart tracts, railways including both meter gauge and broad gauge, tanks, and hillocks, which can be shown on the maps, through their respective conventional signs. The areal distributions of different elements on these maps are clearly shown by using different conventional colouring, shading, and tinting.
- *Easy to understand:* As conventional signs, colours, and shades are used, maps look attractive and informative and can be easily comprehendible by any map reader.
- *Easy to depict the data:* The most decided advantage of drawing these maps is that cumbersome calculations for the gathered data are unnecessary. These maps are qualitative and non-quantitative in nature.
- *Minute Variation:* Chorochromatic maps can be used to show minute differences in the regional distribution of several elements in any region by use of tints and symbols which can be used for regional planners.

Problem: Draw a Chorochromatic Map of Land-Use of the Shanrpur Mouza; P.S. – Ghatshila from the given data.

Table 1: *Data Table for Drawing of Chorochromatic Map*

1. H	20. P	39. P	58. Pond (W)	77. P	96. P	115. P	134. H	153. V	172. H	191. V
2. H	21. H	40. P	59. V	78. P	97. P	116. P	135. H	154. H	173. P	192. V
3. V	22. V	41. P	60. V	79. H	98. P	117. P	136. H	155. H	174. P	193. V
4. H	23. V	42. P	61. RL	80. V	99. P	118. P	137. H	156. H	175. V	194. V
5. R	24. P	43. P	62. H	81. R	100. P	119. P	138. H	157. H	176. P	195. V
6. V	25. P	44. P	63. H	82. W	101. P	120. H	139. H	158. V	177. P	196. P
7. H	26. P	45. P	64. H	83. V	102. P	121. H	140. H	159. H	178. H	197. P
8. R	27. P	46. H	6.5. H	84. H	103. R	122. H	141. V	160. H	179. P	198. P
9. H	28. P	47. PP	66. H	85. H	104. R	123. H	142. V	161. V	180. H	199. V
10. H	29. P	48. H	67. R	86. H	105. V	124. P	143.V	162.V	181. P	200. V
11. H	30. P	49. H	68. P	87. R	106. V	125. P	144. P	163. V	182. P	201. V
12. RL	31. P	50. H	69. V	88. V	107. P	126. P	145. V	164. V	183. P	202. V
13. P	32. P	51. P	70. H	89. H	108. P	127. P	146. H	165. V	184. P	203. V
14. P	33. P	52. P	71. G	90. V	109. P	128. P	147. H	166. Nullar (W)	185. P	204. V
15. P	34. P	53. H	72. P	91. H	110. P	129. P	148. H	167. P	186. P	205. V
16. V	35. P	54. V	73. V	92. V	111. H	130. V	149. H	168. V	187. P	206. V
17. P	36. P	55. H	74. V	93. H	112. R	131. V	150. H	169. V	188. P	207. V
18. P	31. P	56. G	77. P	94. H	113. H	132. R	151. V	170. P	189. W	208. V
19. P	38. P	57. G	76. P	95. H	114. H	133. V	152. V	171. V	190. Mill (H)	
1. H	20. P	39. P	58. Pond (W)	77. P	96. P	115. P	134. H	153. V	172. H	191. V
2. H	21. H	40. P	59. V	78. P	97. P	116. P	135. H	154. H	173. P	192. V
3. V	22. V	41. P	60. V	79. H	98. P	117. P	136. H	155. H	174. P	193. V
4. H	23. V	42. P	61. RL	80. V	99. P	118. P	137. H	156. H	175. V	194. V
5. R	24. P	43. P	62. H	81. R	100. P	119. P	138. H	157. H	176. P	195. V
6. V	25. P	44. P	63. H	82. W	101. P	120. H	139. H	158. V	177. P	196. P
7. H	26. P	45. P	64. H	83. V	102. P	121. H	140. H	159. H	178. H	197. P
8. R	27. P	46. H	6.5. H	84. H	103. R	122. H	141. V	160. H	179. P	198. P
9. H	28. P	47. PP	66. H	85. H	104. R	123. H	142. V	161. V	180. H	199. V
10. H	29. P	48. H	67. R	86. H	105. V	124. P	143.V	162.V	181. P	200. V
11. H	30. P	49. H	68. P	87. R	106. V	125. P	144. P	163. V	182. P	201. V
12. RL	31. P	50. H	69. V	88. V	107. P	126. P	145. V	164. V	183. P	202. V
13. P	32. P	51. P	70. H	89. H	108. P	127. P	146. H	165. V	184. P	203. V
14. P	33. P	52. P	71. G	90. V	109. P	128. P	147. H	166. Nullar (W)	185. P	204. V
15. P	34. P	53. H	72. P	91. H	110. P	129. P	148. H	167. P	186. P	205. V
16. V	35. P	54. V	73. V	92. V	111.H	130. V	149. H	168. V	187. P	206. V
17. P	36. P	55. H	74. V	93. H	112. R	131. V	150. H	169. V	188. P	207. V
18. P	31. P	56. G	77. P	94. H	113. H	132. R	151. V	170. P	189. W	208. V
19. P	38. P	57. G	76. P	95. H	114. H	133. V	152. V	171. V	190. Mill (H)	

KEYS : H - House, V- Vacant, P – Paddy, R – Road, RL – Railway, PP – Petrol Pump, G – Garden, W – Water body

Chorochromatic Map
Illustrating Land-Use Pattern of Shanrpura Mouza P.S. Ghatshila

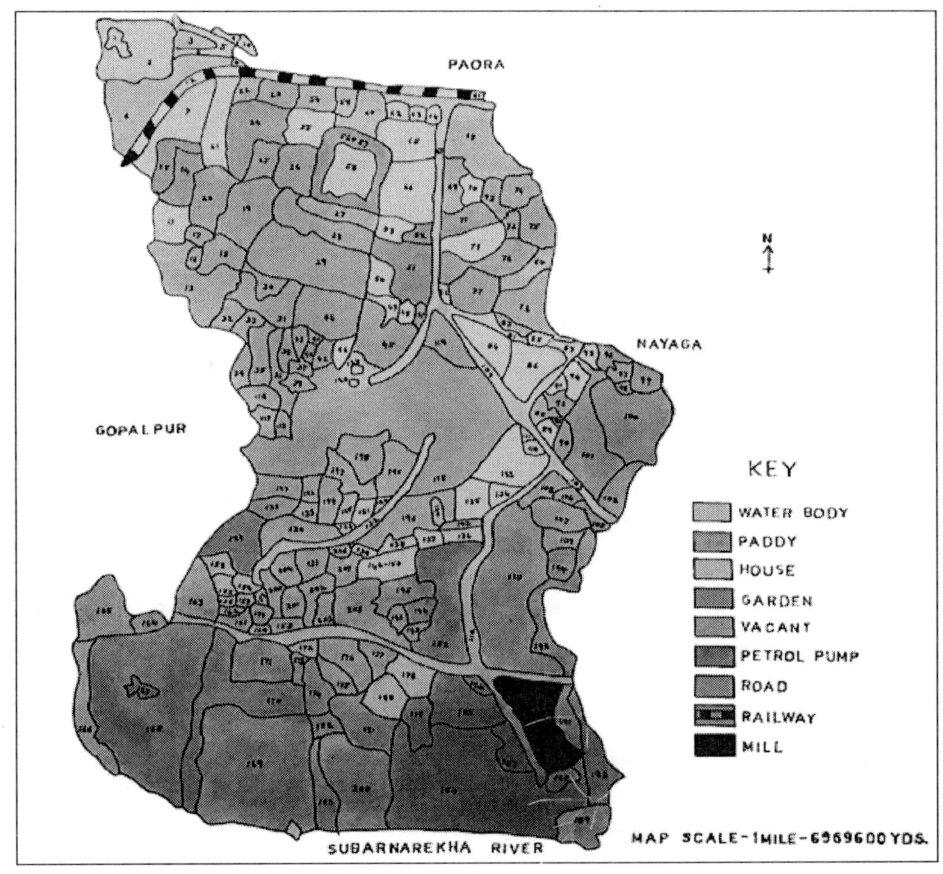

8

Weather Map

Introduction

This weather map of Sunday 16th May 1954 was observed at 17:30 hrs. I.S.T. (1200 hrs. G.M.T.). It is seen that summer prevails over the Indian sub-continent.

Temperature

With the increase in the heat of the land, the cold weather season changes to warmer and warmer weather. The high pressure of the north-western sub-continent which is the characteristic of the cold weather season gradually, changes to the low pressure of the hot weather. In summer the average temperature is about 32° to 33° C. From the isotherms drawn it can be stated that the temperature is about (32° + 8°) = 40 °C. In the southern states of India, low temperature prevails due to the maritime influence.

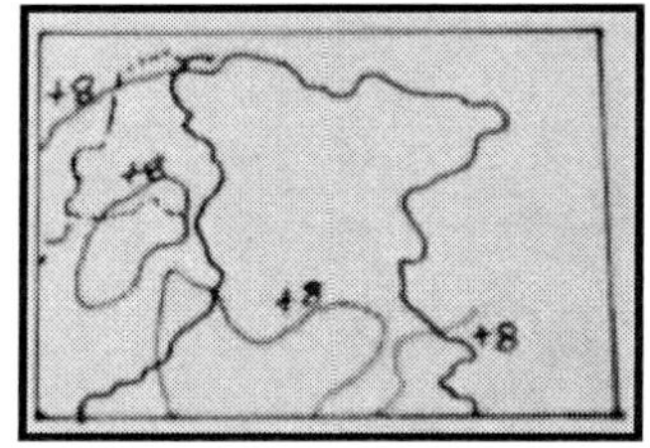

30° N — 35° N
70° E — 80° E

A rough hand sketch

Pressure

Due to the rising temperature in the interior landmass, the pressure is low in the 994 mb, 996mb, and 998 mb isobars are predominant. Comparatively, in the south, 1006 mb, and 1004 mb isobars are seen due to the lesser high temperature prevailing there.

Another important trend of the isobars is that it runs parallel to the border of southern India. There is a low-pressure trough in central India – due to the formation of a local thunderstorm. Moreover, low pressure exists in the north-western frontier provinces due to the western disturbances. These cellular-like pressure zones in a region are termed as bar high when high pressure exists and bar low when low pressure exists.

Another important feature is that there is a steep pressure gradient in the south-western coastal region because isobars are narrowly spaced. Moreover, a steep pressure gradient exists over the land. Whereas, isobars are widely spaced on the sea which

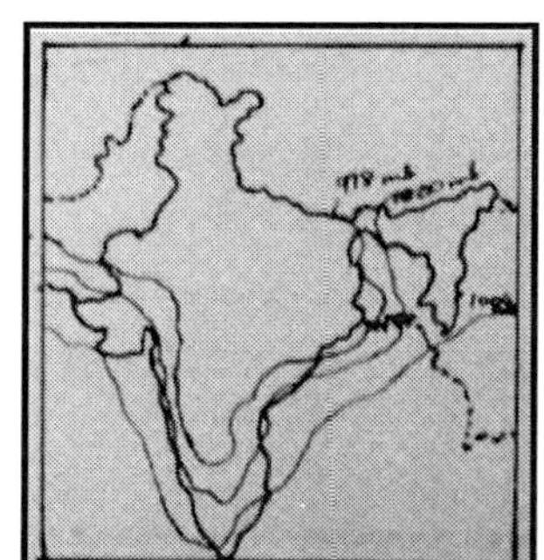

37° 6' N — 8° 4' N
68° 7' E — 97° 25' E

A rough hand sketch

results in a gentle gradient. But in the Sind, Punjab, and Baluchistan portions the isobars are relatively closer and therefore the pressure is quite moderate.

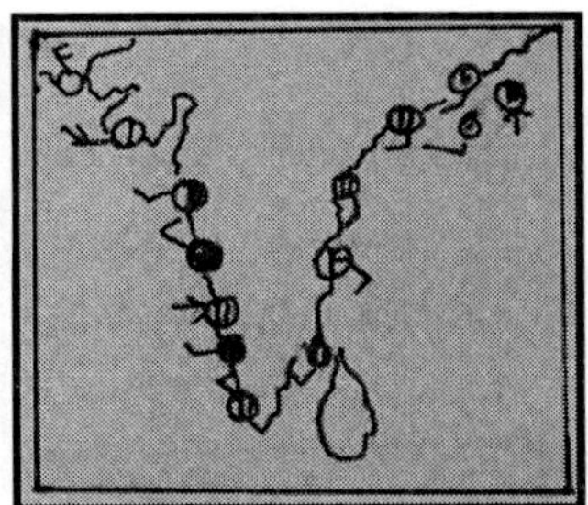

5° N — 22° N
67° E — 90° E

A rough hand sketch

Wind

As high pressure exists over the sea and low pressure on the land, wind direction is obviously directed from the sea towards the land. In some places, due to local disturbances and intense heating, the wind-direction changes. Velocity primarily depends on the gradient of pressure. Since the gradient of pressure was gentle the wind velocity almost throughout the middle sub-continent was also less. Calm to light conditions prevails over the west and north-west India, Uttar Pradesh, Bihar, Bengal, and Assam. But, on the western coast, due to the steep gradient, there is a higher velocity than 50 knots.

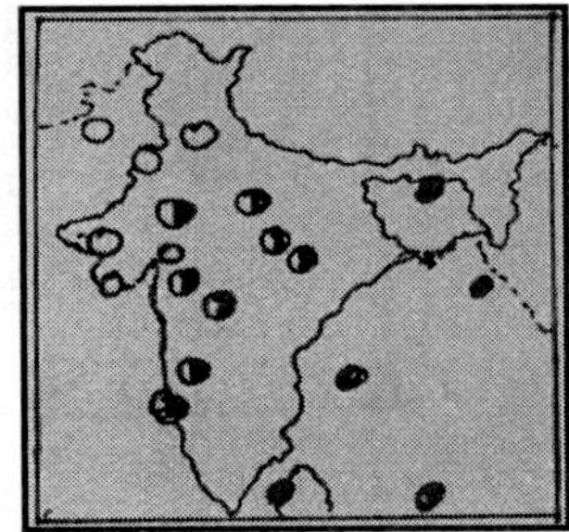

37° 6' N — 8° 4' N
68° 7' E — 97° 25' E

Cloud Cover

In northern India, the sky was almost clear of clouds except locally, at a few places, where the cloud varied from $^1/_8$ to ½. One notable exception was eastern India where at Dibrugarh and Tezpur the sky was overcast. But in the south, mainly in the western coastal plain, the sky was more than ½ clouded.

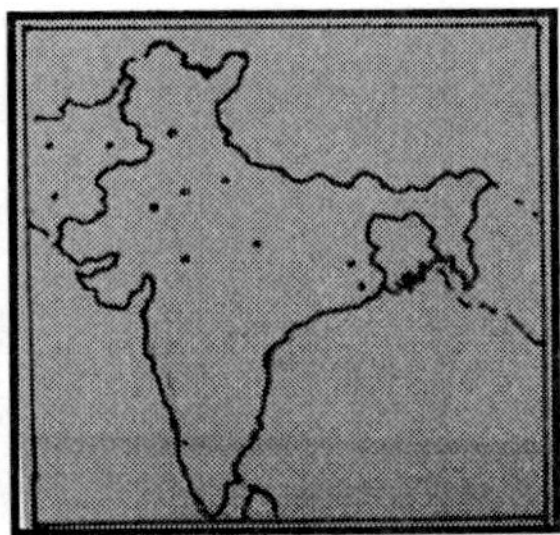

37° 6' N — 8° 4' N
68° 7' E — 97° 25' E

High or medium clouds prevailed over Assam, and in Madhya Pradesh. The Peninsular had chiefly high or medium clouds except in the western coastal plain.

There is no fog or mist which is recorded. But thunderstorms are reported in the area near Ceylon.

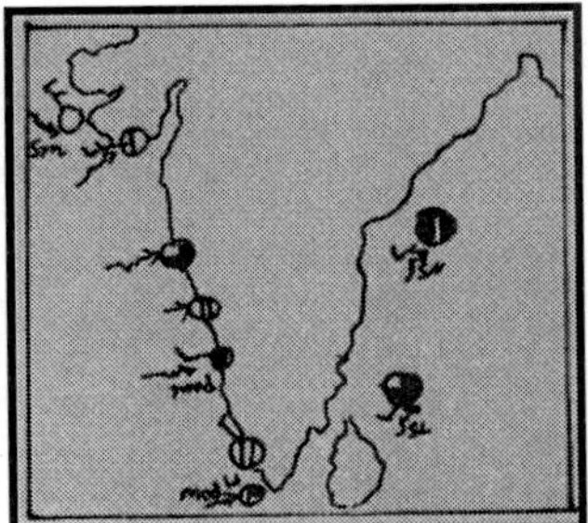

5° N — 22° N
67° E — 90° E

A rough hand sketch

Precipitation

Due to the intense heating and high temperature, there is no possibility of

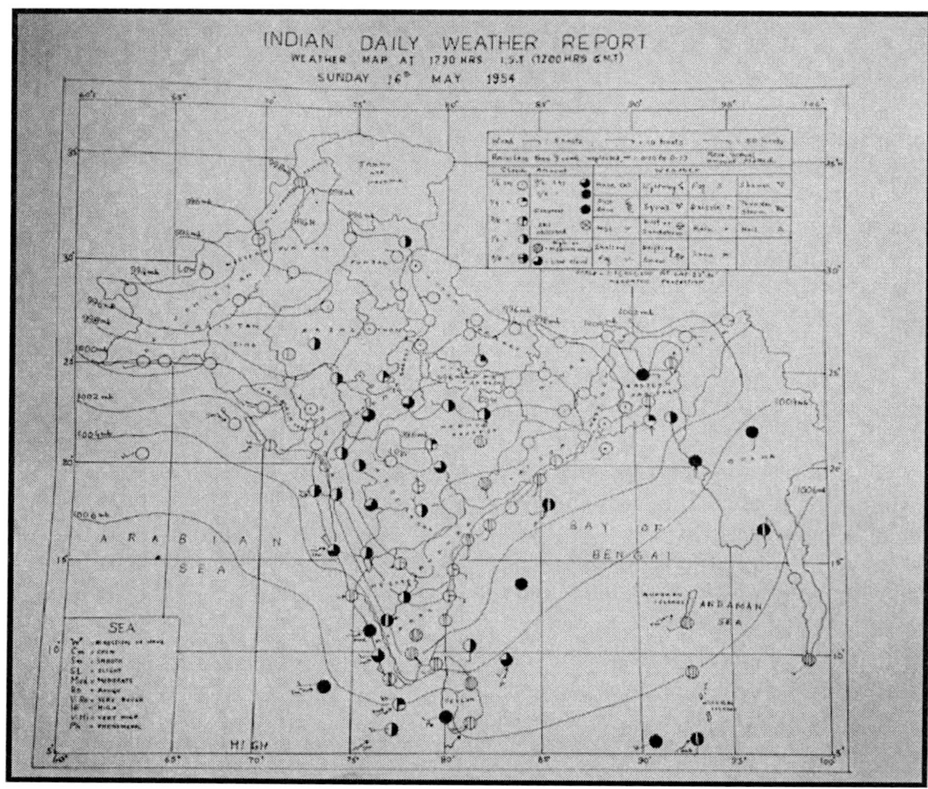

rainfall. Only local thunderstorms in the form of 'Kalbaishakhi' bring rainfall to give relief from the hot dry unbeatable summer.

Sea Conditions

The sea is mostly calm with moderate waves prevailing over it. The waves are directed from the south-south-west to the north-east. As the waves are moderate, the sea is naturally calm and not turbulent.

9

Map Projections

Cylindrical Equal-Area Projection

Introduction: The cylindrical equal-area projection has been derived by projecting the surface of the globe with parallel rays on a cylinder touching it all the equator. The circles of latitudes and longitudes both are projected as straight lives intersecting, one another at right angles. The area between two parallels is made equal to the corresponding surface on the sphere at the cost of great distortion in shape towards higher latitudes. This is why it is an equal-area projection.

Graphical Construction: A circle from the centre O with the radius of the reduced sphere is drawn. OE is produced to Q making EQ equal to the true length of the equator. The tangent TM at E is drawn which should be parallel to PP', from the middle of EQ, draw the central meridian NS right-angled to the EQ, Oa', Ob', Oc', and Od' are drawn in accordance to the given interval. From a', b', c' lines are drawn parallel to the line EQ. All the meridians should be parallel to the tangent, equal in length, and according to the interval given. In this way, the TGLM required graticule is obtained (Refer Figure 1).

Problem: Prepare a graticule for Africa on a scale of 1:50,000,000 at an interval of 10 with latitudinal extent – 40° N to 40° S and longitudinal extent – 20° W to 60° E.

Properties: The scale along the Equator and other parallels become increasingly exaggerated and along meridians, it is progressively diminished. This leads to pronounced deformation of shape in high latitudes. This projection is suitable for Equatorial regions and for low latitudes, where both shape and area are more or less correctly represented.

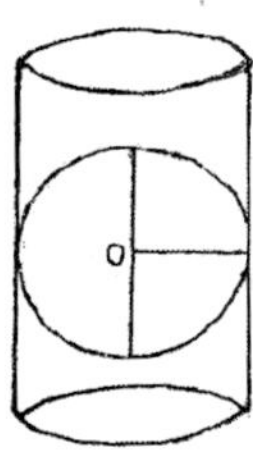

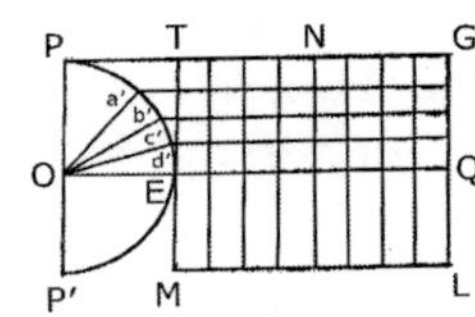

TGLM is the required graticule

Figure 1.

Simple Conical Projection with One Standard Parallel

Introduction: In simple conical projection with one standard parallel the cone is supposed to touch the sphere along the Central Meridian which is truly represented by the graticules. Other parallels are drawn at the true distances from it but the scale along these points is joined with the common centre to show the meridians.

Graphical Construction: The reduced radius is calculated and a circle is drawn. 10 and 50 angles are marked off with the help of a protractor.

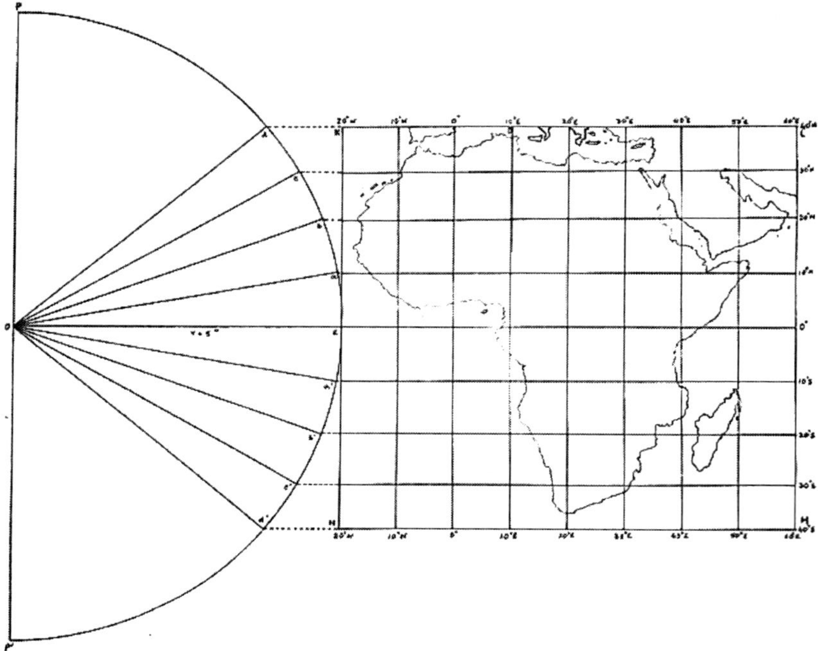

Figure 2: *Cylindrical Equal–Area Projection*

A tangent AV to the circle is drawn at A so VA is the projected radius of the 50 N latitude. ∠rOQ = 10° is the given interval between the latitudes. Therefore, Qr is the true distance between the two parallels at 10° intervals. Now, with centre 0 and radius Qr a semi-circle is drawn which bisects OA at b. From b, a line parallel to OQ is drawn, and the line VA meeting OV at point a is also drawn. Thus, ab is the longitudinal distance between two meridians at an interval of 10° along the standard parallel. Then VO in the centre of the paper is drawn with centre V and radius VA an arc of 50° N is drawn. From the standard parallel, points for the parallels are marked along the central meridian. Similarly, along the arc of 50° N, the distance 'ab' is marked for the spacing of meridians is done. Straight lines from V passing through these points are drawn and thus, the complete graticule is obtained (Refer Figure 3).

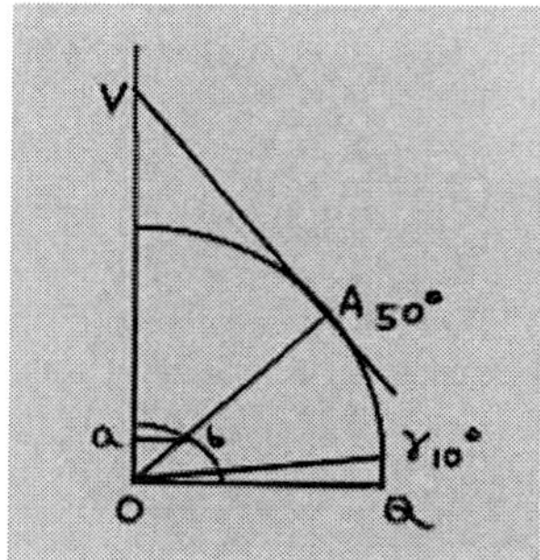

Figure 3.

Problem: Prepare a graticule for Europe on a scale of 1:70,000,000 at an interval of 10° with latitudinal extent – 10° N to 90° N and longitudinal extent – 170° W to 50° W.

Properties: The parallels are concentric circles and the meridians are straight lines and have the correct scale along them. The projection may be completed for one

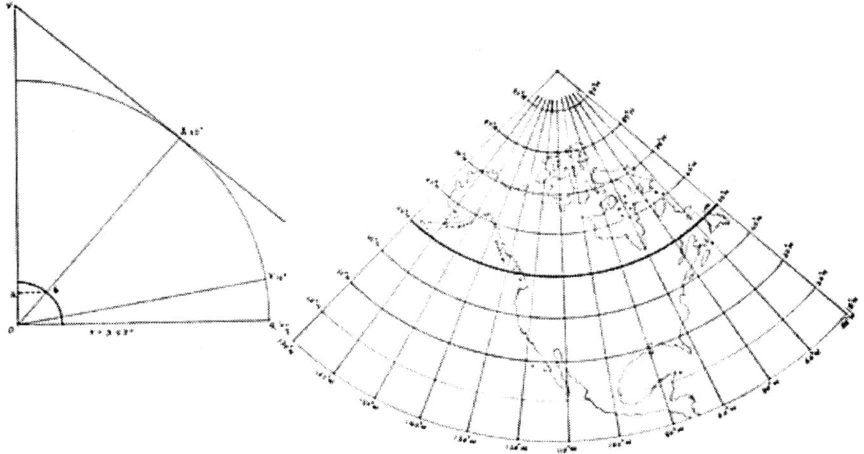

Figure 4: *Simple Conical Projection with One Standard Parallel*

complete hemisphere but suitable for temperate mid-latitude regions, which can be accurately represented.

Polar Zenithal Equal Area Projection

Introduction: This projection was designed by J.H. Lambert in 1772. Like other zenithal projections in this system also the meridians are straight lines, drawn at true angular distances radiating from the pole and parallels are concentric circles. The circles of latitudes become closer away from the pole. Their spacing is so adjusted as to make it an equal area projection. This is commonly used for polar areas.

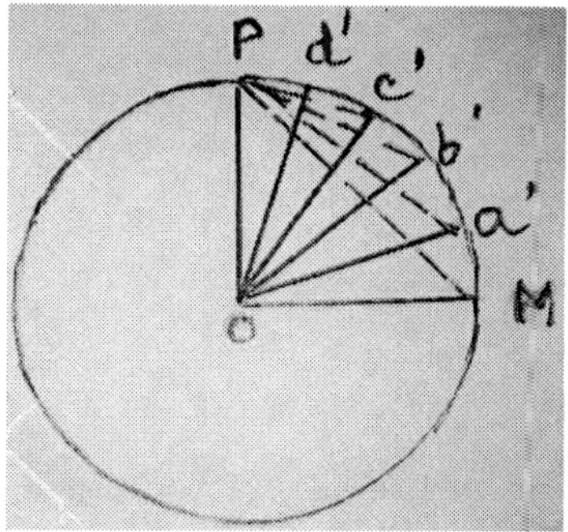

Figure 5.

Graphical Construction: With centre O and with the measurement of the reduced sphere, a circle is drawn to represent the reduced sphere. Through O, a vertical line to meet the circumference at P in the north is drawn to represent the polar point of the reduced sphere. At 15° interval, the angles are marked off from the Equator to the Pole. P is joined to the point of intersection of the arms of the angles of the circumference (Refer Figure 5).

With the centre O and radii each equal to these radiating lines circles to represent the different latitude from 0° to 75° N is drawn. Through 0°, a vertical line / Central Meridian is drawn and on either side of it points at 15 intervals are marked. From the centre to these points, radiating lines are drawn to represent the longitudes.

Problem: Prepare gratitude for Arctic Region on a scale of 1:125,000,000 at an interval of 15° with latitudinal extent - 0° to 90° N and longitudinal extent of 180° E to 180° W.

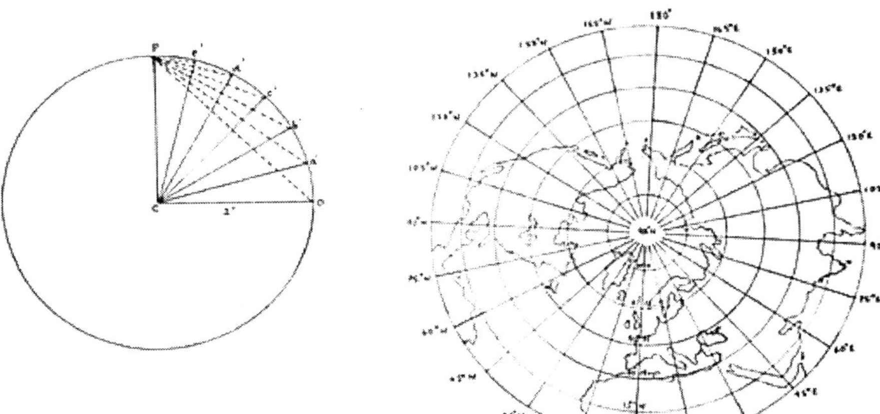

Figure 6: *Polar Zenithal Equal–Area Projection*

Properties: This projection is suitable for polar regions, but distances are observed at the edge of the map. The meridians show the correct bearing or azimuths of all the points from the centre of the map. As the area is maintained, there are distortions in the shape of the continents. But in this projection, the world can be represented in hemispheres only, with the pole as the centre.

10

Topographical Maps

Topographical maps, also known as general-purpose maps, are drawn at relatively large scales. These maps show important natural or physical and cultural or non-physical features such as relief, vegetation, water bodies, cultivated land, settlements, and transportation networks, etc. Other regional maps are drawn by National Atlas and Thematic Mapping Organisation (NATMO).

The Survey of India (SOI) prepares the topographical maps of India for the entire country. These topographical maps are drawn in the form of a series of maps at different scales. Hence, in the given series, all maps employ the same reference point, scale, projection, conventional signs, symbols, and colours.

The topographical maps in India are prepared in two series, i.e.

1. India and Adjacent Countries Series and
2. The International Map Series of the World.

1. *India and Adjacent Countries Series:* Topographical maps under India and Adjacent Countries Series were prepared by the Survey of India till the coming into existence of adjoining countries was abandoned and the Survey of India confined itself to preparing and publishing the topographical maps for India as per the specifications laid down for the International Map Series of the World. However, the Survey of India for the topographical maps under the new series retained the numbering system and the layout plan of abandoned India and Adjacent Countries Series.

   The topographical maps of India are prepared on 1:10,00,000, 1:250,000, 1:1,25,000, 1:50,000, and 1:25,000 scales providing a latitudinal and longitudinal coverage of 4°×4°, 1°×1°, 30′×30′, 15′×15′, 5′×7′30″ respectively. The numbering system of each one of these topographical maps is shown in Figure 1.
2. *The International Map Series of the World:* Topographical maps under the International Map Series of the World are designed to produce standardized maps for the entire World on a scale of 1:10, 00,000, and 1:250,000.

Reading of Topographical Maps

The study of topographical maps is simple. It requires the learners to get acquainted with the legend, conventional signs, and colours on the sheets. The conventional signs and symbols depicted on the topographical sheets are shown in Figure 5. The whole of India and adjacent countries bounded by Latitude 0° to 40°N and Longitude 60°E to 100°E were subdivided into uniform sections of 4°×4° dimensions from 1 to 137, implying that the fundamental basis of these topographical maps are the well-defined latitudes and longitudes of the Earth.

These 4°×4° topographical sheets are drawn on Lamberts Conical Orthomorphic Projection to the scale of 1:10, 00,000 (or 1inch to 16 miles) and are also known as 'Million Sheets' or 4′ sheets or 1M Maps. The present political or international boundary organization, there are 137 1M sheets covering all the major parts of India.

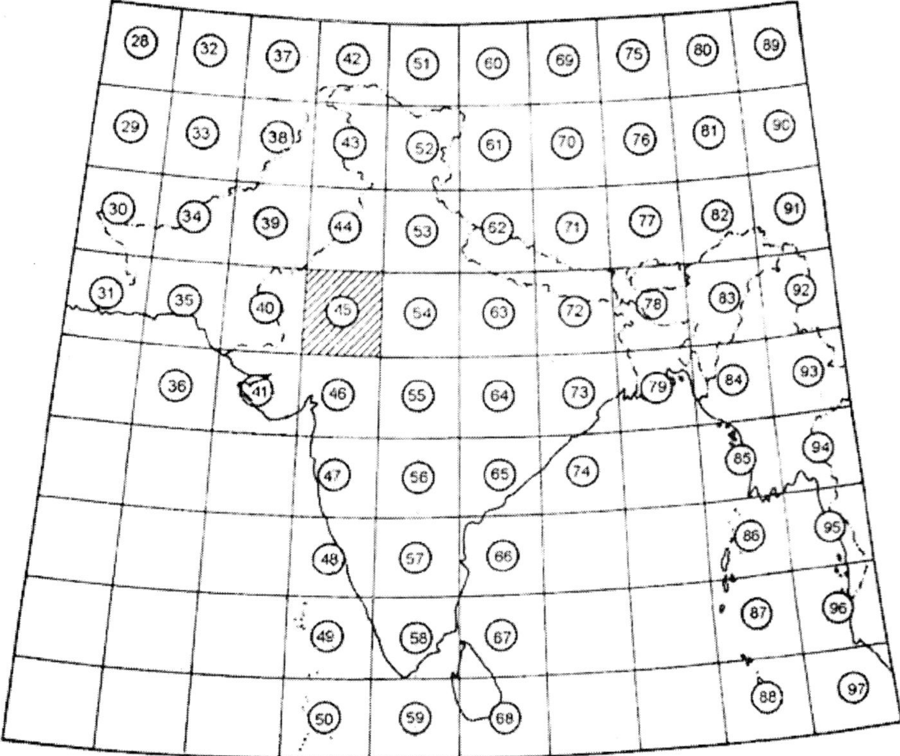

Figure 1: *The Map Series Covering the Indian Subcontinent*

Each of these 1M sheets, Toposheet No. 55 is further divided into 16 uniform divisions of 1°×1° dimensions and are designed by 16 letter symbols of English alphabets from A to P as depicted in Figure 2a. These 1°×1° sheets are drawn to the scale of 1:250,000 (or 1 inch to 4 miles) and are referred to as 'quarter-inch sheets' or 'degree sheets'. (Refer to Table No 1)

Topographical Map 55D/7

Table 1: *The Different Sheets of Indian Topographical Maps (According to Scale)*

Sheets	*Scale (Metric System)*	*Scale (FPS System)*	*Dimensions*	*Contour Intervals*	*Examples*
Million Sheet or 1M Map	1:1,000×10³	1 inch to 16 miles	4°×4°	—	55
Quarter-inch or Degree Sheet	1:250×10³	1 inch to 4 miles	1°×1°	100m/250ft/500ft	55D
Half degree or Half inch Sheet	1:100×10³	1 inch to 2 miles	30′×30′	50m/100ft/250ft	55D/NW
Quadrant Sheet or Inch Sheet	1:50×10³	1 inch to 1 miles	15′×15′	20m/50ft/100ft	55D/7
Special	1:25×10³	—	5′×7′30″	10m	55D/7/5

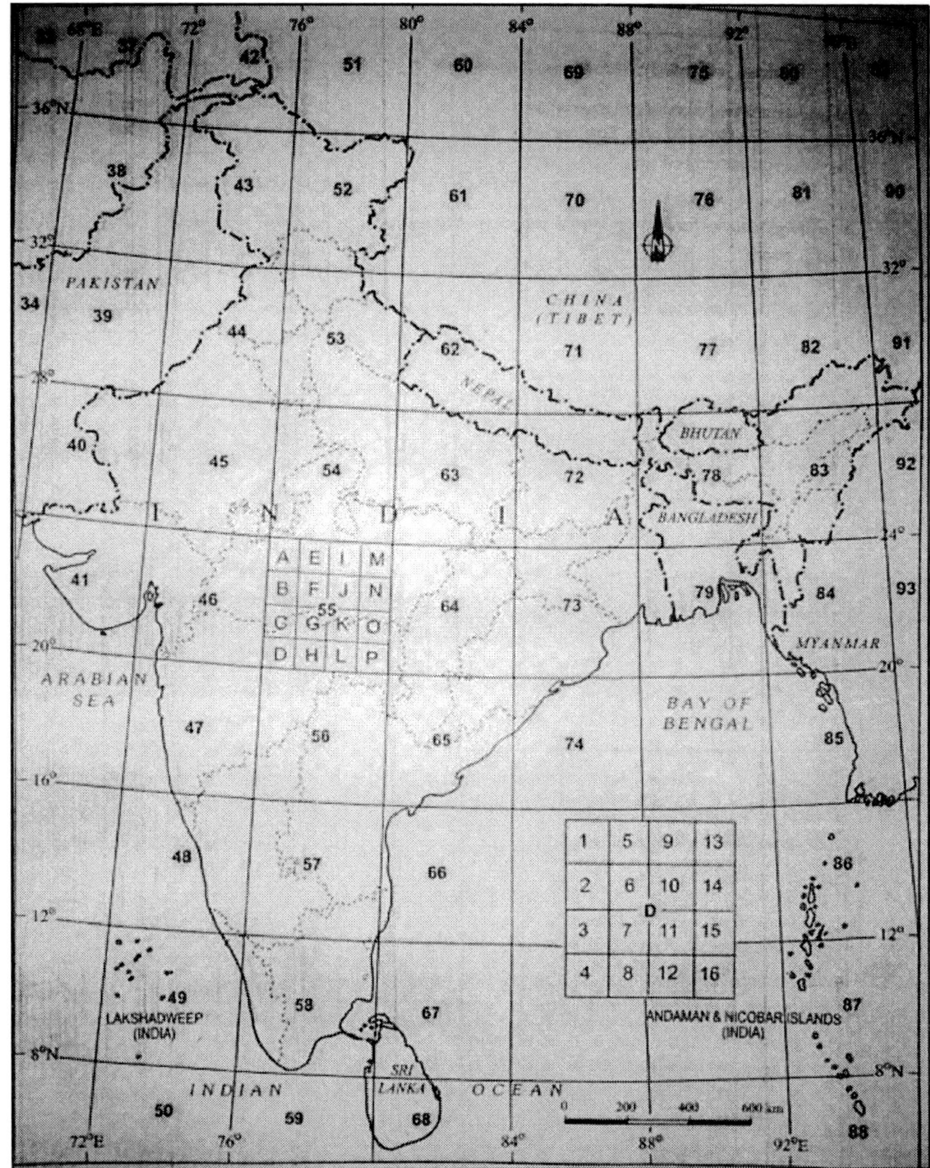

Figure 2a: *Reference of Topographical Sheets Published by Survey of India (SOI)*

Further, each of these degree sheets, for example, D is subdivided into 16 parts of $15'\times15'$ dimensions and is numbered from 1 to 16. These squares are drawn to a scale of 1inch to 1mile, commonly marked as 1:50,000, but being 1:63,360 to be exact in the ratio form. Each of the squares is identified by an index number; for example, the one shown hatched in Figure 2b, has the index number 55D/7.

Each of the 1:50,000 sheets is further subdivided into four 1:25,000 scale sheets of $7^{1}/_{2}'\times7^{1}/_{2}'$ dimensions and the four squares are labeled NW, NE, SW, SE as shown

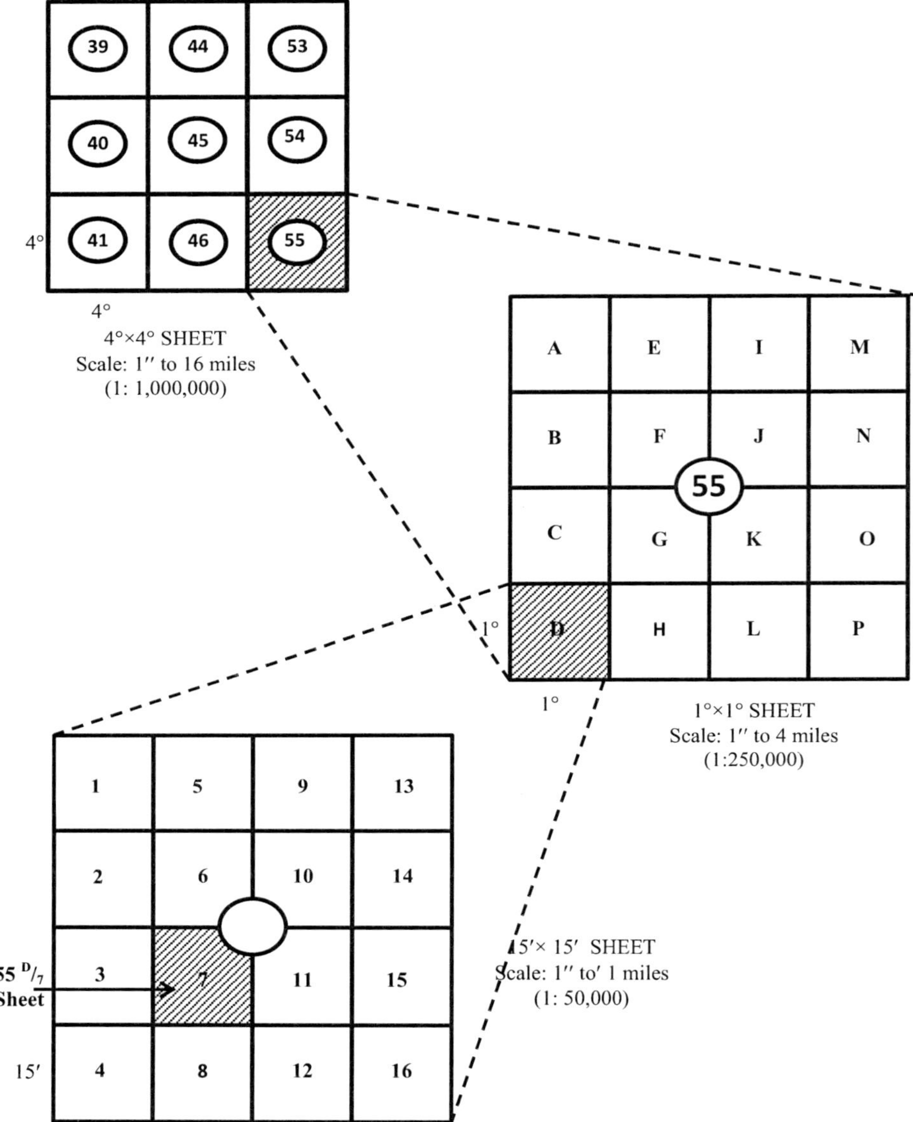

Figure 2b: *The Different Subdivision of the Square 45 with the Classification of the Scale*

in Figure 3a, and identified appropriately. Alternatively, the sheet is divided into 6 divisions (5′ latitude × $7^1/_2$′ longitude) and numbered 1 to 6 as depicted in Figure 3b.

Grid Reference

The topographical maps bear the national grid of squares drawn to the scale of 2cm = 1 km, with each square having a side measuring 1 km. These are the sheets most commonly used for various purposes and are of vital practical importance.

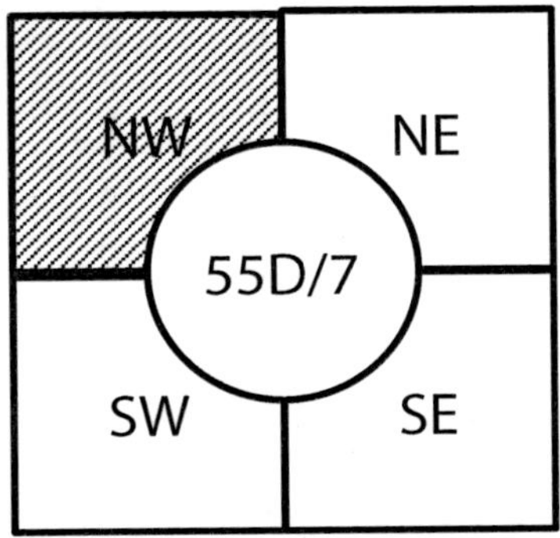

Figure 3a: *1:25,000 Topographical Maps (7½′ latitude×7½′ longitude)*

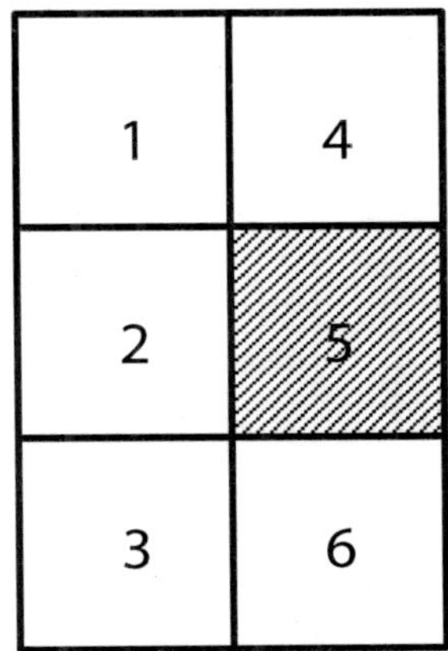

Figure 3b: *Old-layout Topographical Maps (5′ latitude×7½′ longitude)*

Figure 3a and 3b: *The Detailed Division of a Square*

With reference to any of these sheets, the vertical lines are called Eastings since they are numbered from West to East. Similarly, the horizontal lines are called Northings because they are numbered from South to North (Figure 4.)

This network of horizontal and vertical lines, or the Eastings and Northings, is called the Grid. It is also identified as the Grid Reference.

Principles of Map Interpretation

The topographical map gives a representation of the area on paper, using conventional signs and symbols, in order to describe the area of land use represented on the map. The objective of map interpretation is to bring out the relationship between various physical and cultural features.

Physical Features: The contour lines provide the basis for the study of the relief features of an area. The location of the highest point and lowest point and their height depict the general slope of the area which can also be inferred from the direction of major rivers in the area. Contrasts in slope may be examined with reference to the spacing of contour lines and physical features as valley or lowland, upland, ridge, hill, cliffs, rock outcrops, sand hill, ravines, etc. may be inferred.

Drainage Features: The main river and the directions in which they flow, whether the rivers are perennial or not, and the presence of waterfalls, rapids, etc., may also be noted. The presence of dams, canals, and distributaries may indicate that the rivers are used for irrigation.

The presence of non-perennial streams shows that the region probably receives low rainfall. The location of tanks, wells, springs, marshes, and swamps, and other drainage features bring out the relation between the relief and drainage features.

Land Use: The type of natural vegetation such as forest, grass, or scrub gives an indication of the climatic conditions of the area. The relation between physical features and land use may also be examined. As, the area probably gets low rainfall the distribution of wastelands, and sparse vegetation with palmyra and cactus is widespread. A large area

Topographical Map

State: West Bengal — Toposheet No.:73 $^{M}/_{10}$

Districts: Barddhaman and Birbhum

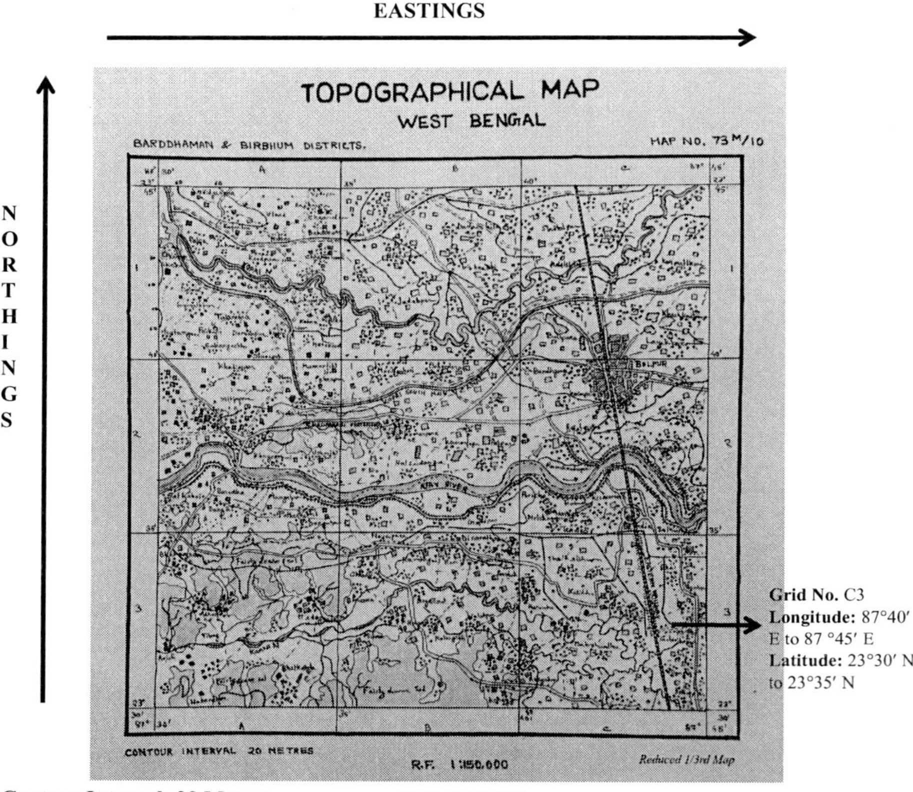

Contour Interval: 20 Meters **R.F. 1:50,000**

Figure 4: *Grid Reference of the Topographical Map*

of dense forest may indicate that the area probably receives high rainfall. Cultivated land is likely to be found to a large extent in lowlands and river valleys, whereas, regions of comparatively high relief and steep slopes are likely to have forest or scrub jungle and they are not suitable for cultivation.

Settlements: The average distance between villages in different parts of the area signifies the settlement to be nucleated or dispersed. Settlements are likely to be more closely spaced in the valleys and lowlands than in the uplands, plateaus, and hilly areas. The site of settlements may be on the banks of rivers, near tank bunds, or near wells. These settlements are called *wet-point settlements* as they are near the sources of water supply. In such cases, one may infer that the region probably received low rainfall, and settlements seek sources of water for their location. In low-lying areas such as the flood plains of rivers or in the deltas, settlements are located on higher ground above the level of floods and may be located on river terraces, higher mounds, or near the foothills away from river courses. Such settlements are called *dry-point settlements.* Some settlements are located on hilltops or near forts. These are *strong-point* settlements.

In some areas, typical patterns of settlements may also be found. Linear patterns of settlements are found along roads, river banks, levee ridges, etc. Nucleated, square, or rectangular patterns of settlements may occur at the junction of roads. Dispersed settlements may be found in highly dissected forest regions, plateaus, coastal area and the areas vulnerable to floods.

There may be evidences of urban or suburban growth, nodal centres with transport routes and other amenities. It may be of square, rectangle, star-like, or circular shapes. The deserted village sites may be in forested regions or flood-prone sites.

Transport and Communication: The railway lines may be broad gauge or metre gauge; single track or double track and there may be cuttings, embankments, or tunnels along the railway line. Railway lines are not found in the low-lying zones of the land, rugged relief features, or in areas of sparse population.

The patterns of roads in the area may be metalled or non-metalled. Many settlements are linked by cart tracks or footpaths, bridges, or causeways across rivers which may have ferry services. The route pattern of roads and railway lines are related to the physical features of the area.

Occupation of the People: The occupations of the people may be inferred directly and indirectly from the evidence regarding land use and settlements. If the area has large tracts of cultivated land, the dominant occupation is likely to be agriculture which may be emphasised by irrigation from tanks, wells, or canals and cultivation is likely to be of the intensive type with wet crops like rice and sugar cane, being the common ones. If the area does not exhibit any evidence of any source of irrigation, cultivation is likely to depend on rainfall only. Places of quarries or mines are likely to provide occupation to the people. In the case of villages only along the coast, fishing is likely to be important. Large cities are likely to provide employment in industries, trade and services. These may be brick kilns, textile or rice mills, carpet factory, cement factory, etc. In the case of towns which are headquarters of districts and *taluks,* a large number of offices, administrative jobs, and other related services of the tertiary sector may be marked on the map which provides employment.

Transect Chart

Usually, *Transect Charts* are drawn to present the correlation between the physical and cultural elements present in a particular topographical area. An ideal section line is taken cutting through all the elements. This section line represents almost all the physical and non-physical (cultural) elements possibly (maximum extent) present in the map (may be rural or urban based).

Usually, it is within a frame of 5 to 6 tiers. The lowest tier marks the relief features (relief typicals present) with undulations in accordance with the cross-section taken from the topographical map. The second tier comprises the drainage network (both perennial, and non-perennial rivers), water bodies, and broken banks of rivers. The third tier depicts the natural (RF and PF) and planted vegetation with species-types. The fourth tier presents the transport (road, railways, water or ropeways or may be airways) and communication network (Post Office, Telegraph Office, Post and Telegraph Office,

Police Station). Minor transport routes like cart tracks and pack tracks in rural areas need to be considered. The top-most tier represents the settlement characteristics, types, and patterns. These may be around favourable sites of low relief, transport linkage, availability of water, cleared vegetation, agricultural land and other infrastructural amenities. The maps of the older edition had settlements (black in colour) whereas the newer editions have settlements (red in color).

These elements are arranged and designed in a frame so as to draw a correlation between all the above-mentioned elements and also to draw the inference about the climate, economy and, occupational pattern. Moreover, the problems and prospects of the region can also be identified. This process of identification would enable the planner to bring about regional development in the area by the initiation of several beneficial projects.

Precautions need to be taken so that the main river channel has no overlaps of vegetation, communication transport linkages and, settlement points in their respective tiers.

Conventional Signs or Symbols Used in Topographical Maps

The different geographical elements are described with different signs in topographical maps. So, to study the topographical sheets it is very important to have a concept about such conventional signs. These signs are developed and used by the Survey of India. Such signs have the following characteristics:

(a) Conventional signs are not drawn to the scale.
(b) The signs are easily understandable.
(c) In some special cases signs are selected in such a way that a separate explanation is not needed.
(d) Use of colour is very important like blue for water bodies, yellow for agricultural area, green for vegetation cover etc.

To interpret the topographical maps the following signs and symbols are to be studied:

1. (i) Metalled road, according to importance, with milestone, (ii) unmetalled road, according to importance, with milestone. 2. Cart-track 3. Pack-track 4. Footpath with bridge 5. Pontoon bridge 6. Ferry or Ford 7. Railway and roadway on river-bridge 8. (i) Roads in dry bed, (ii) Unlined Stream 9. Canal 10. Dam 11. Big stream—(i) perennial, (ii) sandy bed, (iii) waterline in sandy bed, (iv) riverbed with islands and rocks. 12. Tidal creek 13. Waterfalls with altitude 14. Rapid 15. Steep stream bank with height and broken bank 16. Submerged boulders 17. Shallow submerged bank 18. Swamp or Marsh 19. Reeds 20. Well—(i) lines, (ii) unlined. 21. Tube-well 22. Spring 23. Tank—(i) perennial, (ii) fluctuating, (iii) perennial with an embankment. 24. Excavated tank—(i) perennial, (ii) fluctuating, (iii) perennial with an embankment. 25. Embankment along roads 26. (i) Unmetalled road on an embankment along tank, (ii) unmetalled road beside embankment along tank 27. Broad-gauged railways—(i) double line, (ii) single line with stations, (iii) under construction. 28. Other railway- (i) double line, (ii) single line with stations (iii) under construction. 29. Tramways 30. Railway under tunnel 31. Railway station with siding 32. Telegraph line 33. Level crossing 34. Contours—(i) lines in shape, (ii) stony waste (iii) cliff. 35. Sand dunes—(i) flat, (ii) heaped, surveyed, (iii) transportational 36. Village—(i) permanent, (ii) relict. 37. Huts—(i) permanent,

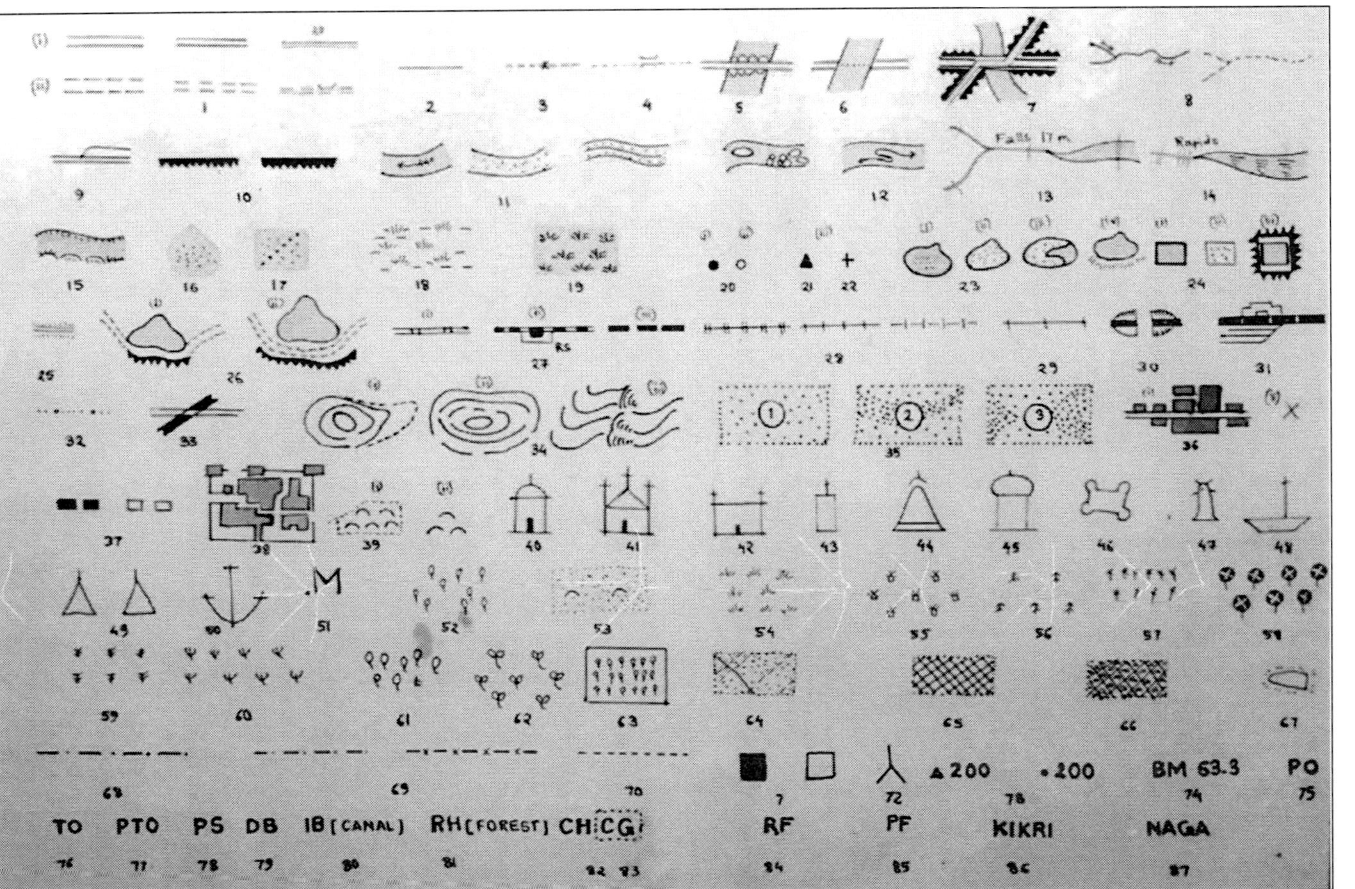

Figure 5: *Conventional Signs or Symbols Used in Topographical Maps*

(ii) temporary. 38. Village with walls 39. (i) Graveyard, (ii) Graves 40. Temple 41. Mosque 42. Idgah 43. Church 44. Pagoda 45. Tomb 46. Fort 47. Lighthouse 48. Lightship 49. Buoy—lighted, without light 50. Anchor 51. Mines 52. Trees 53. Shrubs 54. Grasses 55. Bet 56. Coniferous tree 57. Palmyra trees 58. Palms 59. Bamboos 60. Cactus 61. Other trees 62. Banana trees 63. Gardens 64. Tea garden 65. Betel 66. Vegetables 67. R.F. boundary 68. International boundary 69. State boundary 70. District boundary 71. Boundary pillar- (i) Surveyed, (ii) Unsurveyed. 72. Triangular point of village. 73. Heights surveyed 74. Benchmark- 63.3m 75. Post Office 76. Telegraph Office 77. Post and Telegraph Office 78. Police Station 79. Dak Bungalow 80. Inspection Bungalow 81. Rest house 82. Chimney 83. Circuit house 84. Reserved Forest 85. Protected Forest 86. KIKRI (Tribes) 87. Naga (Tribes).

A few specimens of topographical map interpretations of the plateau and plain regions have been discussed in the following section.

Topographical Map Interpretation (Map No.1)

(This map represents features of a plateau geographical area)

Introduction

1.	Topographical sheet number	:	73 $^{A}/_{7}$
2.	State	:	Part of Bihar
3.	Districts	:	Parts of Palamau and Ranchi
4.	Extent	:	Latitude- 23°15′ N to 23°30′ N, Longitude-84°15′ E to 84°30′E.
5.	Scale	:	1 : 50,000 (2 cm to 1 km)
6.	Contour Interval	:	20m
7.	Area	:	703.8 sq.km
8.	Surveyed in	:	1979-80
9.	Published in	:	1982
10.	Surveyor	:	Major General Girish Chandra Agarwal, Surveyor General of India

From the primary observation, it is revealed that the region has two distinct subdivisions- the northern portion represents a dissected plateau while the southern portion represents an erosional plain.

Relief

The relief of the map area is dominated by high plateau region which is eroded by the rivers, therefore making it a dissected plateau region. The general slope of the land is from south to north along the direction of river flow. The map area is sloping towards the river Koel which flows through the middle of the area. The highest peak is 1068m in the west-central part of the map (Grid A_2). Though the river valley has the lowest relief, yet it is 600m. The region can be broadly divided into three sectors based on physiography. They are discussed in the following section-

Above 900m (hilly region): The entire western part and east-central part of the map falls under this category with average height of 980m. Though this is a plateau region, many eroded hills are found here, like conical hill, flat top hill, low ridges etc. along with a number of widened ‘v’-shaped valleys.

Figure 6: *Topographical Map of 73 $^{A}/_{7}$*

700-900m (higher plateau region): The region is mainly found in the west, east and central part of the area along with the middle part of the river valley. Though the average height is 900m, the region is sloping from the high hills towards the river Koel. In some places cliffs are noticed while in other parts gentle slopes along with Cols and Saddles are also found.

Below 700m (river valley region): This region has gentle slopes, but on the basis of height this can still be considered as a lower plateau region with average height of 650m. This region has a general slope from south to north.

Drainage

The main river of the area is Koel, flowing from south to north direction in the central part of the map. The river valley, though narrow, it is quite deep. It has divided the area

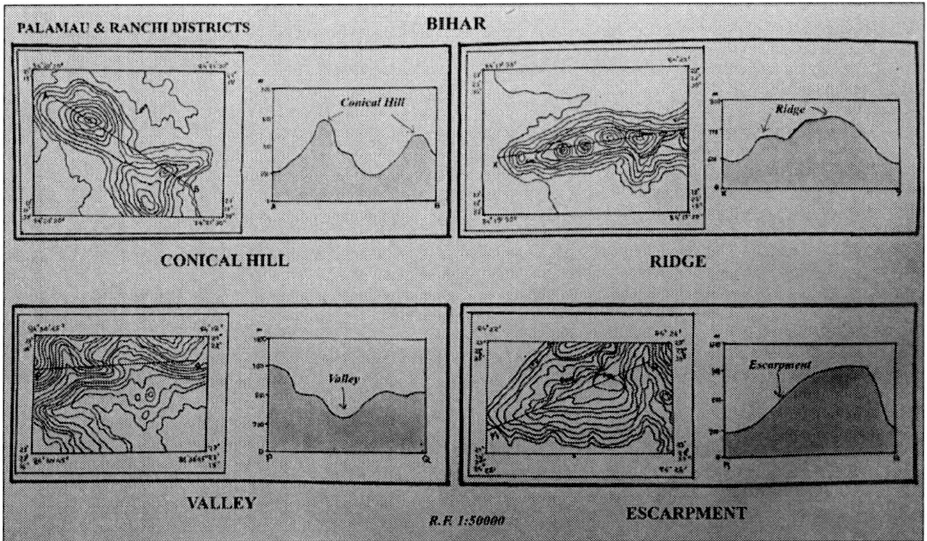

Figure 7: *Different Relief Features of the Topographical Map of 73 A/7*

into two halves. Though perennial, huge amounts of sand and boulder are deposited in the river bed. The river is fed by numerous small streams on both sides. A big tributary Phuljhar river joins Koel in the central part of the area. Like Koel, Phuljhar also has a meandering course but it is non-perennial with the deep valley. Many drainage patterns are created by the small stream network throughout the area.

Drainage Patterns

Dendritic Drainage Pattern: It mainly occurs in those regions with uniform lithology. Here the alignment of tributaries and sub-tributaries are analogous to branching arrangement of the tree with a confluence angle typically acute. It is thus a well-developed drainage system with subsequent, consequent and obsequent streams.

The map area depicts a number of dendritic drainage patterns, all over it. A typical example of it is found near the Bishnupur settlement at 23°25′8′′ N to 23°26′6′′ N and 84°24′94′′ E to 82°24′4′′ E.

Parallel Drainage Pattern: It has developed due to alternate beds of hard and soft rocks. This is mostly found in areas having closely spaced counters, in high slope areas. In the map it is found at the extreme south- western corner near the Deoragang P.F. at 23°27′33′′ N to 23°27′57′′ N and 84°10′45′′ E to 84°20′ E.

Radial Drainage Pattern: It is mainly found in places where conical hill or dome-like structures are present from where the streams radiate out in different directions and is mainly controlled by relief. The most prominent example of radial drainage is at 23°27′33′′ N to 23°27′28′′ N and 84°22′9′′ E to 84°25′6′′ E.

Ring-like Pattern: It is mainly developed in places where hill and dome-like structures are present. The streams flow roundly of the hills and makes a ring-like, pattern. The most prominent example of ring- line drainage is found at 23°17′55′′ N to 23°19′ N and 84°15′ E to 84°17′ E.

Most of the rivers, here, form waterfalls. A large numbers of channel features like river-meander, badland topography and braided channels have also developed.

Natural Vegetations

About 80 per cent of the entire map area is covered by forest. Both protected and reserved forests are found here. The notable protected forests are Chorkakhar P.F., Barang P.F., Manata P.F., Porha P.F., Birmala P.F., Balato P.F., Sato P.F., Tahdar P.F., Saraideh P.F., Nindi P.F. etc. Parts of this Ranchi and Palamau districts also have reserved forests of which the most noteworthy are Rema R.F., Ranga R.F., Musir R.F. and the Katia R.F. Most of these forests are fairly dense with mainly the sal trees. There are also some open mixed jungles as well as some open scrubs or dense scrubs. However, plantations have been made in almost everywhere.

In some regions forests have been cleared for building up human settlements like parts of Menne P.F. Most of the forests are drained by a large number of tributaries of the North Koel river.

Settlements

The term 'settlement' refers to the establishment of inhabitation in a particular region. A number of settlements have been found in the Palamau and Ranchi districts of Bihar. As the area is quite rugged with thick forest cover, the settlements

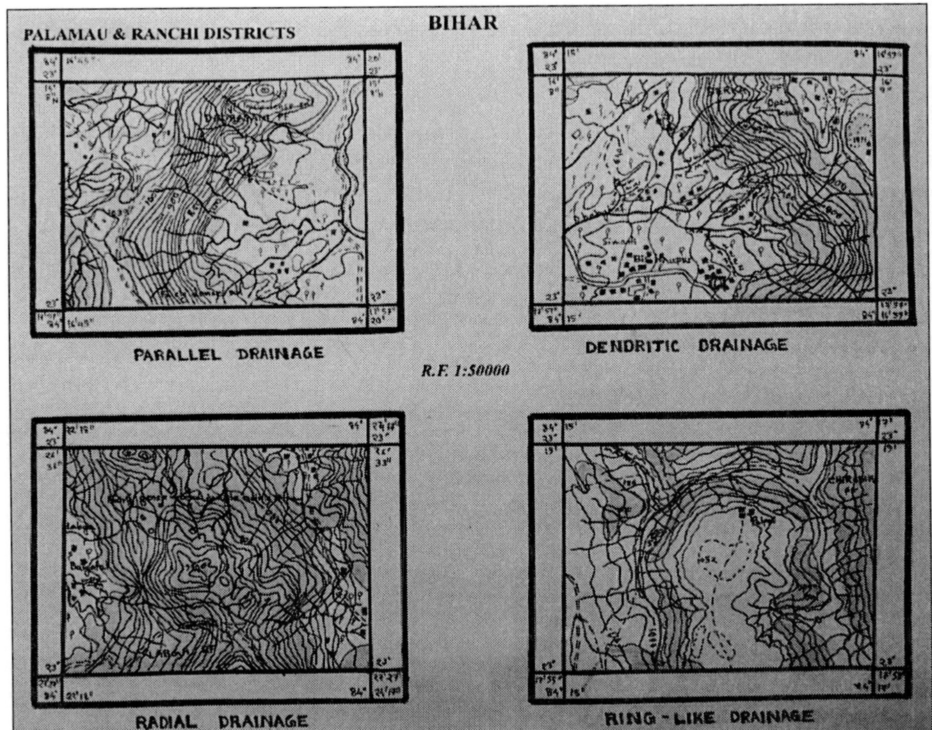

Figure 8: *Different Drainage Pattern of the Topographical Map of 73 A/7*

are not well developed here. Different types of settlements are found. These are mainly of three types:

Nucleated Settlements

Nucleated settlements are found at the junction of metalled roads and river. They develop in a region where many modern day civic amenities are present like post and telegraph office, electric sub-station, hospital, tanks. The communication network facilitates the growth of settlement.

Such a settlement is found in the area extending from 84°22′12′′ E to 84°23′57′′ E and 23°21′33′′ N to 23°22′28′′ N. The settlement is known as the Bishnupur settlement and has the amenities like overhead tank, dispensary, market, post office and telegraph office, police station and electric line and telephone line.

Linear Settlements

Linear settlements are usually developed along some important roadways or by the banks of the rivers. Such a settlement is found in the Palamau and ranchi districts of Bihar extending from 84°18′48′′ E to 84°19′20′′ E and at 23°26′51′′ N.

Isolated Settlements

Isolated settlements emerge as scattered hamlets which are set up surrounding to certain basic civic amenities like post office, market etc. Such a type of settlement is found in a latitudinal extension of 23°13′14′′ N to 23°30′ N and a longitudinal extension of 84°26′53′′ E to 84°30′ E. There are three settlements namely Jurni, Asurtali, Pirtaul which are located amidst the Birtau P.F.

Transport and Communication

Typically, the region does not possess a well-developed transport network. It is because there are extensive tracts of reserved and protected forests hindering the transport development. Mainly all kinds of roads *i.e.* both metalled and non-metalled are found in this area. The only broadest metalled road of the map stretches from north-western part *i.e.* Neterhat to south-eastern part of the map to Ranchi. The road covers a distance of about 100 km.

The area is also criss-crossed by a number of unmetalled road segments, which are usually motorable in dry season. It has been found that the metalled road extends from Neterhat, which is the focal point of traffic in this region to Ranchi. There are a lot of pack-tracks which serve as the connecting point between two villages. Hence transport development as a whole remained at a poor level except in some places like on levee, flood-plain margins of river North Koel and also on valley plains of some tributaries. The general patterns are –

(i) The valley of river North Koel is as such better served by roads.
(ii) The north-eastern part of the area has no means of communication except pack-track. Other parts enjoy more or less, better transport facilities.

Conclusion

The map area is not well-developed with proper infrastructural facilities due to the ruggedness of topography and extensive dense forest. The transect chart, a correlation technique, has been used to show the relationship between the physical and cultural features of a region. It has been found that in regions of high relief there is dense

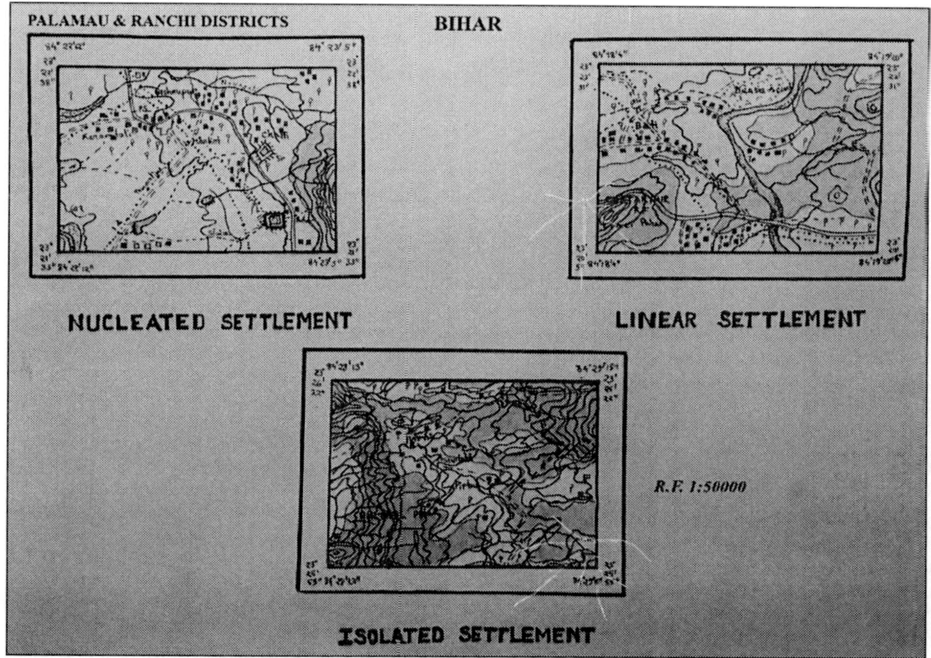

Figure 9: *Different Transport, Communication and Settlement Patterns of the Topographical Map of 73 $^{A}/_{7}$*

vegetation as the forests are not cleared to prepare agricultural fields in the regions. Moreover, these high relief zones have fewer roads and fewer settlements. The presence of comparatively non-resistant rock masses and the numerous drainage network have led to the denudation of the land surface and lowering of the gradient of the land. Only in these areas, the settlement with transport and communication system has developed. In the near future, this region too can develop with proper land-use planning.

Topographical Map Interpretation (Map No. 2)

(Only Relief and Drainage features have been discussed for this map)

Introduction

1.	Topographical sheet number	:	$73^{B}/_{5}$
2.	State	:	Bihar and Madhya Pradesh
3.	Districts	:	Ranchi and Raigarh
4.	Extent	:	Latitude- 22°45′ N to 23°0′ N, Longitude-84°15′ E to 84°30′E.
5.	Scale	:	1 : 50,000 (2 cm to 1 km)
6.	Contour Interval	:	20m
7.	Area	:	703.8 sq.km
8.	Surveyed in	:	1978-79
9.	Published in	:	1980
10.	Surveyor	:	Kishori Lal Khoshla, Surveyor General of India

The general relief of the region represents dissected plateau topography. In the southern part, the relief is a high plateau region, where the highest contour is 784m. But the highest point of the region is 940 m which is observed in the eastern and north-eastern parts of the region. In the northern region, scattered high hills with a height above 700m are found in this region. The slope of the map is from the north to the south. In the western part, the highest height is 840m. There is a decline in slope from west to the middle of the map and from east to the middle of the map because of the

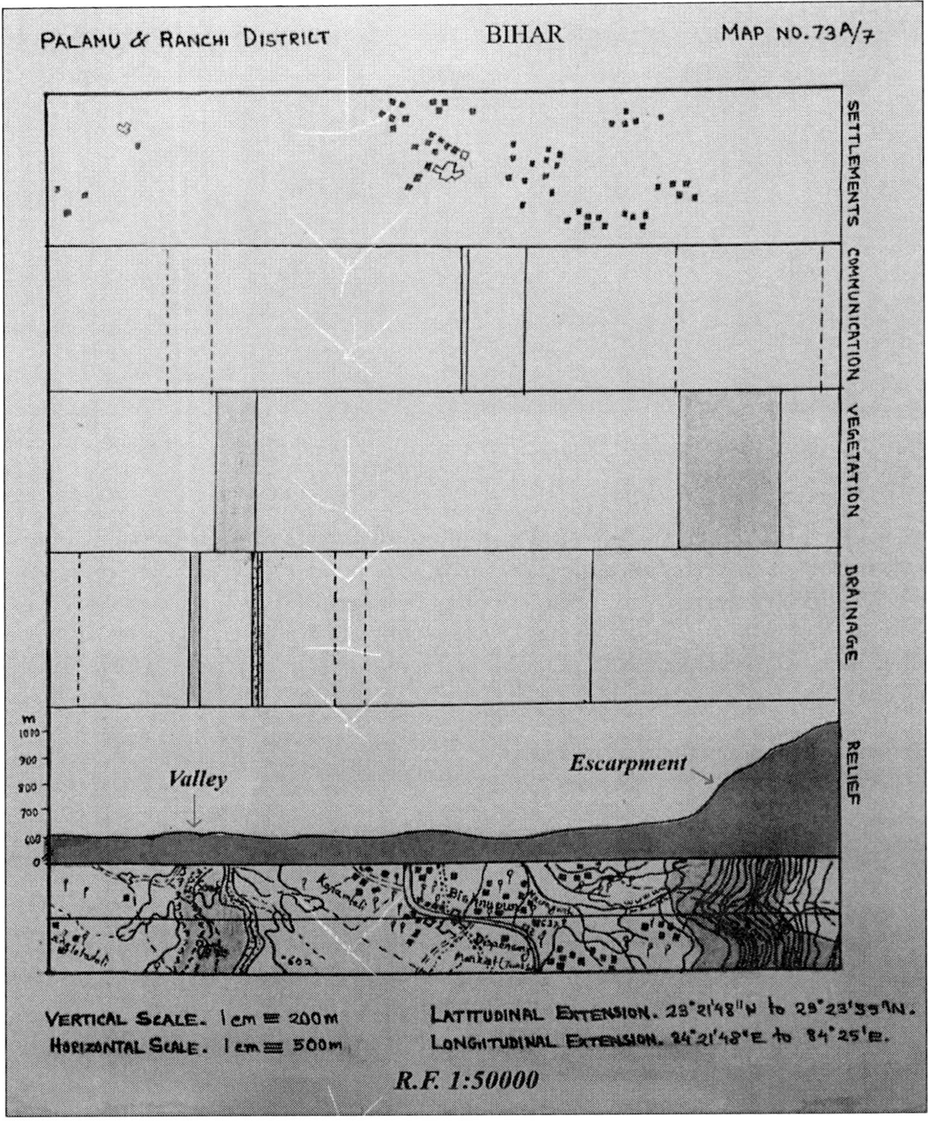

Figure 10: *Transect Chart depicting Interrelationship among Physical and Cultural Features of the Selected Area of the Topographical Map of 73 A/7*

river flowing through the middle of the region. But as a whole, the general slope is from north towards south. So it can be said the river Sankh dissected the region into two parts—eastern and western. In the north-eastern part, the highest relief is 893m and lowest contour is 740m. The lowest contour is found 440m in the southern region. The lowest elevation of the relief along the middle of the region has been observed because of the erosion of River Sankh. So many cliffs have been found in the southernmost part of hill-slopes. The dense drainage network made the region undulated in topography. The badland topography has resulted because of the erosion of numerous dip streams. A number of conical hills, dome-shaped hills, 'V'-shaped valleys, flat topped hills, ridges and saddles are observed. The stony waste and boulders are also typical features of this region. From all these features it can be said that this region of Ranchi and Raigarh districts is in the part of high plateau region. Some of the relief typical has been discussed in the following section.

Relief Features

Ridges and Saddle: The ridges are the elongated, sloping stretches of high-relief ground area. The middle portion of the ridges is seemed to be narrow and the slope beside the two sides of the narrow high regions is steep in nature. The contours of the ridges are elongated and spaced very close but some highly scattered contours are found in the middle, those mainly indicate the highest peaks of the ridges of resistant rock masses.

Saddles are dips between mountains and hills and along the erects of the ridges, ridge lines and spurs and are so named because they resemble a riding saddle viewed from the side. They can be wide or narrow, shallow or deep.

In the topographical map, the saddle and ridges are found in grid C3 with the extension of 22°53′ N to 52°52′ N and 84°27′30′′ E to 84°30′ E. This depicts simple ridges and saddle as it appears on a topographical map and in profile given. The saddles here are mainly formed because of the erosion of the rivers.

Conical Hills: A conical hill is a hill with a uniform slope and is represented by concentric contours spaced almost regularly. Highest contour value is observed in the middle of the circular contour and the lowest contour is observed in the outermost part of the circular contours. This is the hill that is round at the base and rises up to a sharp peak like a cone.

In the topographical map, the conical hill is found in the northern portion within the extension of 22°57′ N to 22°58′ N and 84°23′20′′ E to 84°25′9′′ E. Here, the highest elevation is 940m.

Sheet Rock and Rocky Knob: The sheet rock are the large area of rocks where the overlying soil layers have been eroded and removed due to mechanical weathering. In the topographical map a sheet rock is found in A3 grid. The highest relief in this portion is 740m.

Rocky knob is totally a barren rock and exposed as a rock exposure. The trails of the rocky knob vary from easy to strenuous.

Cliff: A few small areas of steep slopes have formed cliff like features with contours adjacent to each other. Such cliffs may be related to collapse of the lower part of the slope or to any other structural cause.

In the map, a cliff has been found within the extension from 84°27′ E to 84°28′30′′ E and 22°47′ N to 22°48′ N. The cliff has been formed because of erosion by the dip stream.

Isolated Hills: These hills are found at advanced stage of dissection. An isolated hill may be a monadnock too. Over such hill radial pattern of drainage network develop.

In the topographical map, isolated hills have been observed within the extension of 22°53′ N to 22°54′55′′ N and 84°23′53′′ E to 84°24′50′′ E. The lowest contour is 640m and highest contour value is 720m. This dissection has occurred by the distributaries of Sankh river in grid B2.

Valley: The valley is a low area between hills or mountains, often with a river running through it. The valley of the river may be of different shapes, such as 'I' shaped, 'V' shaped, or wide 'V' shaped.

In the topographical map a widened valley with 'V' shape has been observed. These widened 'V' shaped valleys are formed mainly for the erosion of the river in the youthful stage. The contour valleys are increased from the inner portion toward the outer portion of the valley. The highest contour is 600m and the lower contour is 580m in the grid C3.

Flat-Topped Hills: A flat-topped hill is an elevated area of the land. A flat top and the sides that have usually steep cliffs are known as butte, mesa. The mesa has a top wider than its height and they are all formed by erosion and weathering of horizontally layered rocks. In the given map, the highest contour value of flat-topped hill is 784m situated in the B3 grid extending from 22°46′30′′ N to 22°47′ N and 84°21′30′′ E to 84°22′30′′ E. Buttes, though absent in the map have greater height and top is less wide.

Dome: The irregular contour, and slightly spaced, rounded in shape are the domes. These are not very high in relief. They are also found in an isolated and scattered manner in the given map.

Drainage Patterns

The drainage pattern means the form of the drainage systems and the spatial arrangement of streams in a particular locality or region. Many drainage patterns are created by small streams network throughout the area.

Parallel Drainage Pattern: Parallel drainage pattern comprises numerous rivers which are parallel to each other and follow the regional slope. This pattern is mostly found in the area having closely spaced contours in high slope areas. In the map it is found in the northern part extending, from 22°58'30" N to 22°59'30" N and 84°17'30" E to 84°19' E.

Radial Pattern: Radial drainage pattern also known as centrifugal pattern is formed by the streams which diverge from a central higher point in all directions. It is obvious that dome structures, volcanic cones, conical hills etc. are favourable for the development of the radial pattern. In the map, a radial drainage pattern has been found on a conical hill in the northern part.

Annular Pattern: Annular pattern also known as circular pattern which is formed when the distributaries or the tributaries of the master consequent streams are developed in the form of a circle. The most prominent example of annular pattern is found in the south east corner of toposheet. Such pattern is developed over a mature and dissected dome mountain characterized by series of alternate bands of hard rock and soft rock beds.

Rectangular Pattern: Rectangular pattern shows some degree of resemblance drainage patterns join their master streams almost at right angle. The rectangular pattern is characterized by widely spaced tributaries. This is found in the map area extending 24°51'30" N to 22 °55' N and 84°25' E to 84°26' E.

Pinnate Pattern: Pinnate pattern is developed in a narrow valley flanked by steep ranges. The tributaries originating from the steep sides of the hills and join the master

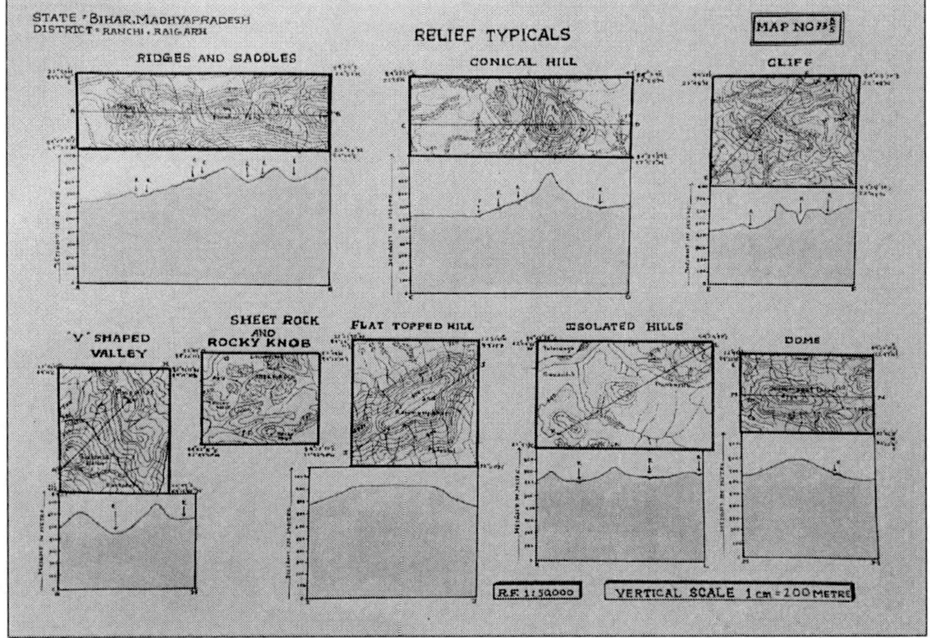

Figure 11: *Different Relief Features of the Topographical Map of 73 $^{B}/_{5}$*

stream in acute angle. This pattern resembles the veins of a leaf. In the southern part a pinnate drainage pattern have found.

Barbed Drainage Pattern: Barbed drainage pattern is developed in a region where the surface in underlying with the hard rocks. This is a rare kind of drainage pattern is formed when the distributaries are flow in opposite direction to their master streams. The tributaries join their master streams in a hook-shaped bend. This type of drainage pattern have found within 22°41' N to 22°50' N and 84°25'40" E to 84°28' E.

Herring Bone Pattern: Herringbone drainage pattern also known as rib-pattern, is developed in mountainous areas broad valleys are flanked by parallel ridges having steep hill side slopes. The longitudinal consequent streams as master streams are developed in the longitudinal parallel valleys while tributaries as lateral consequent after originating hill slopes. Just this type of drainage pattern found in the region extending 84°27' E to 84°28' E and 22°54' N to 22°54'50" N.

Centripetal Pattern: Centripetal or inland drainage pattern is opposite to the radial drainage pattern because it is characterized by the streams which converge at a point which is generally a depression or a basin. This pattern is formed by a series of streams which after emerging from surrounding uplands converge into the basin in the topographical sheet extending 22°53' N to 22°54'40" N to 84°29'40" E to 84°53' E.

Drainage Characteristics

The main river of the area is Sankh river – flowing from north to the southern of the area. This river dissected the whole region into two equal halves. This river is perennial in nature, huge amount of sand and boulder is deposited in the river bed and is fed

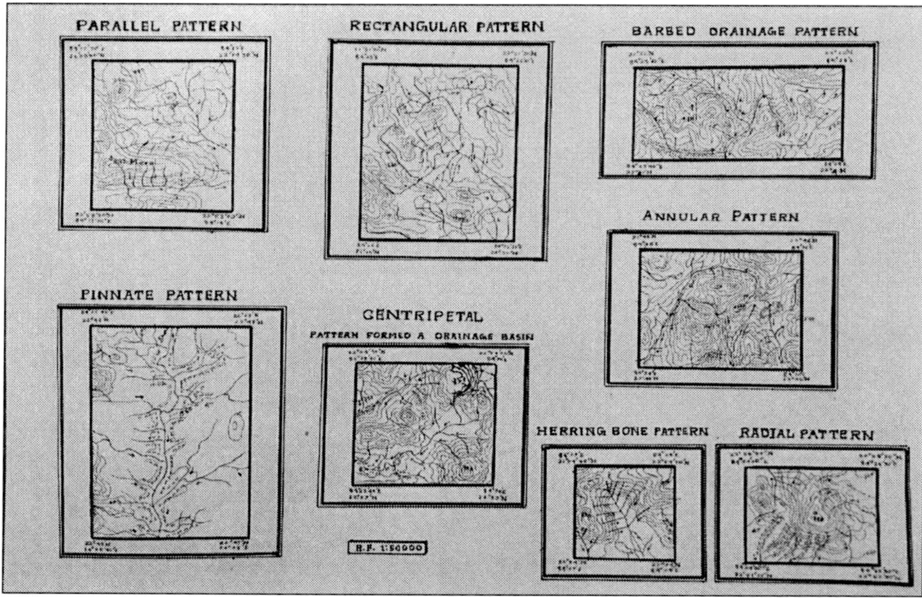

Figure 12: *Different Drainage Pattern of the Topographical Map of 73 B/5*

by numerous small streams on the both sides. A big tributary Bala Nadi joins the Sankh river at the northern part of the area. River Sankh is meandering in nature. The distributaries which join with Sankh in its left side are Milmili Nala, Silphi Nadi and the distributaries which join with Sankh in the right sides are Bahri Nala, Bala Nadi etc. The river is fordable during November to June. In this lower course of the river several riverine features are ideally depicted in the region. They have been discussed in the following section.

Point Bar: A point bar is a depositional landform attached to the bank. In a meandering channel it is crescent shaped. Point bars are formed along the linear banks of the channel in the river Sankh.

Mid Channel Bar: The mid-channel bar has been observed in Sankh river, where the river bifurcates itself into two main parts. These mid-channel bars sometimes lead towards the formation of a braided channel.

Meander Bend: The meandering channel is characterized by pools (deeps) and riffles. In the river Sankh, a meander bar has been observed in loops within the extension of 22°24' N to 22° 30' N and 84°22' E to 84°23'40" E.

Straight Channel: "Truly straight channels scarcely exist outside the laboratory, except where a headwater stream is firmly held on the line of fault" – stated by G.D. Dury, 1969. In the natural system straight channel may be possible up to stretches of 100 meters or so. The straight reaches are generally characterized by riffles at crossovers. In the topographical sheet of Ranchi district the river Sankh is passing through a part of plateau region which formed granite- gneiss and some highly metamorphosed soft rock. So, the southern part of Sankh river became straight channel because of the geology of the area.

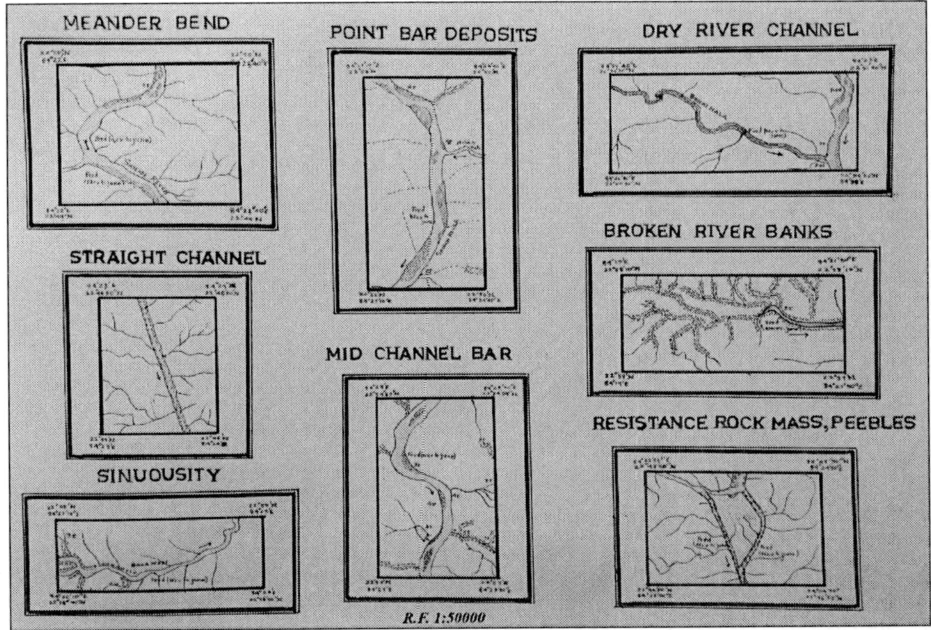

Figure 13: *Different Drainage Characteristics of the Topographical Map of 73 $^{B}/_{5}$*

Sinuosity: The shape of longitudinal segments of river channel is considered in terms of sinuosity which refers to departure of actual channel course from expected theoretical straight line. A channel having the sinuosity of 1.05 to 1.5 is defined as the sinuous channel according to Leopold and Wolman (1957). Sinuosity is the ratio of the river length (distance travelled it one floated down the river) divided by the length of straight river flood (bird's flight distance from source to mouth).

Topographical Map Interpretation (Map No. 3)

(This map represents features of a plain geographical area)

Introduction:

1.	Topographical sheet number	:	$73^{M}/_{10}$
2.	State	:	Part of West Bengal
3.	Districts	:	Parts of Birbhum and Barddhaman
4.	Extent	:	Latitude—23°30′ N to 23°45′ N, Longitude—87°30′ E to 87°45′ E.
5.	Scale	:	1 : 50,000 (2 cm to 1 km)
6.	Contour Interval	:	20m
7.	Area	:	703.8 sq.km
8.	Surveyed in	:	1970-71
9.	Published in	:	1972
10.	Surveyor	:	Dr. Hari Narain, Surveyor General of India

From the primary observation, it is found that the area covered by the map has a great variety of physical and cultural features, and it can be said that it is a plain area.

Figure 14: *Topographical Map of 73 $^{M}/_{10}$*

Relief: The general slope of this area is from west to east as is revealed by the value of contours and from the direction of river flow. The highest points are found in the east-southern portion and the lowest points are found in the west-southern portion. The general value of the contours ranges from 40m-100m. In the northern side, it can be seen that mainly the contours are varying between 40m- 60m. The south-eastern part of the map is consisting of the contour which is below 80m. The highest area in the map is on the northern side. By observing the map it can be said that it is mainly a depositional plain.

Drainage: An overall view of the topographical map shows that the region has one important perennial river and two non-perennial rivers. The main river is Ajay. It flows from western side to eastern side. It can be seen that the river is in its lower stage and have some depositional features on both sides of the banks.

There is the presence of another river named the Kopai Nadi. It flows from the north-western side to north-eastern side. It passes over Monohorpur. This river is perennial and a canal is also there named 'Kopai South Main Canal'.

Kunur Nadi is another river. It flows from south to western zone. Almost half of the year the river remains dry. In this area there are many tanks. In the northern side there is one reservoir near Kanchannagar. Kopai South Main Canal is joined with it.

Natural Vegetation: The vegetation cover in this area is almost absent. The vegetation is mostly restricted to trees planted by man. But some parts of the upper west-southern side have natural forest cover where some fairly dense sal forests have been found. However, there is a protected forest in Chaupahari in the western part of the region. Open scrubs are also found in a small portion of southern side near Warlshpur. There are some reserved forests near Santiniketan.

Settlements: The term 'settlement' refers to the establishment of inhabitation in a particular region. A number of settlements have been found in the Barddhaman and Birbhum districts of West Bengal because of the availability of the favourable factors for establishing settlement. Some settlement patterns have also been identified in this region. They are as follows:

Scattered Settlements: Scattered settlements have developed in the north-western side of the map extending from 23°39′30′′ N to 23°40′ N and from 87°31′ E to 87°32′30′′ E. The settlement like Sakargachho, Khadimpur, Jalanagar are present on unmetalled roads. These are localised near the points of available resources and amenities.

Linear Settlements: The places which are very much prominent for this type of settlement pattern is in the north-eastern side. The metalled road coming from Siuri and the Eastern Railway of Barddhaman Branch is responsible for this type of settlement. These have developed on both sides of the metalled road in C_1 grid between 23°43′30′′ N to 23°44′ N and 87°42′ E to 87°45′ E. This settlement named Sarbanandpur settlement enjoys the facilities of Kopai river, post office, dispensary and other communication facilities etc.

Dispersed Settlements: In the southern side of the map there are dispersed settlement. The houses and buildings are isolated away from each other. It includes Natunnagar, Deolbardh, Hedagarld with a latitudinal extent from 23°32′ N to 23°33′ N and longitudinal extent of 87°30′30′′ E to 87°32′ E. The availability of large tracts of agricultural land and the forested area has resulted in the dispersal of settlements.

Compact Settlements: This type of settlement is seen in the north-western side of the region. The name of the town is Bolpur. It is situated near Ajay river. Power line passes through the area. The site is marked with Post Office, Telegraph Office, Police Station, Visva-Bharati University, Hospital, and Market. This significant settlement is extending from 23°40′15′′ N to 23°41′30′′ N latitudes and from 87°40′ E to 87°42′30′′ E longitudes. There are also some reserved forests in this area.

In the western part of the map, another clustered settlement is found. The name of the places are Islamabad and Bhedia Nagar. They are also situated near Ajay river and Eastern Railway. The presence of PTO PWD Office, Police Station and Market (which opens every Monday and Friday) make the place important.

Transport and Communication: The transport and communication are through railway, roadway and motorway. The following discussion can give a brief idea about the modes of transport in the area.

- Railways: Eastern Railway is mainly found in this area. The railway line connects from north to southern side (to Barddhaman at a distance of 32 km) from Ahmadpur. There are four railway stations *i.e.* Prantik R.S., Bolpur R.S., Pickri and Dihal R.S. There are also some foot-bridges.
- Roadways: There is a national highway in the south-eastern side of the area. Two roads coming from Durgapur and Pivlganga meet with the highway and make a junction in Ilambazar and goes forward Kesarmath. Now from Bolpur, the road makes straight towards southern side. Regarding communication, a well-established network has been introduced through Telephone lines also.
- Waterways: The rivers are fordable during October to June which facilitates cheap connectivity.

Conclusion: The map is of a developing area of the Barddhaman and Birbhum districts. The transect chart shows the relationship between the different physical and cultural features. The relief varies from 45m to 60m. It has been drawn with a latitudinal and longitudinal extent of 23°36′30′′ N to 23°36′ N and 87°35′30′′ E to 87°37′30′′ E respectively. There is an important river *i.e.* Ajay and some non-perennial streams also. The Chaupahari Protected Forest is in this area. There are settlements like Gopalpur, Karnagar, Shichendrapur developed beside the railway and metalled road. These transport routes add special importance to the development of the area.

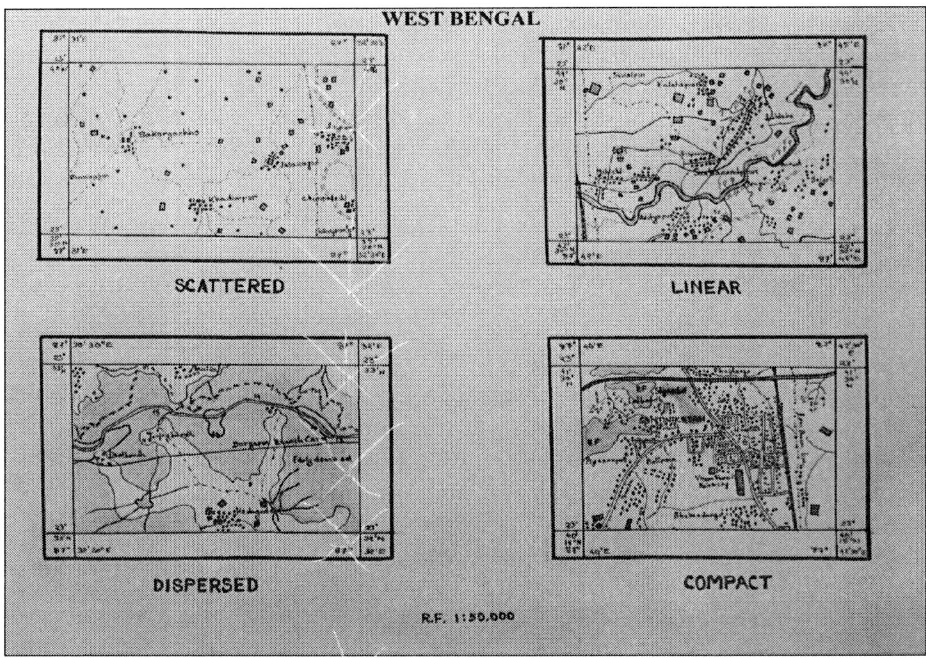

Figure 15: *Different Settlement Patterns of the Topographical Map of* $73^{M}/_{10}$

Due to the lack of fertile agricultural tracts and industrial establishments, the growth and prospect of settlements in this region are very feeble. The local inhabitants have mainly settled around hospitals, market places, post offices from where they can avail services and communicate with others. Man has constantly endeavored to make the region suitable for their living. Thus, the advancement of science and technology and the fruitful implementation in the betterment of practical life will definitely help to surpass the hindrance caused by lacking resources for the development of the region.

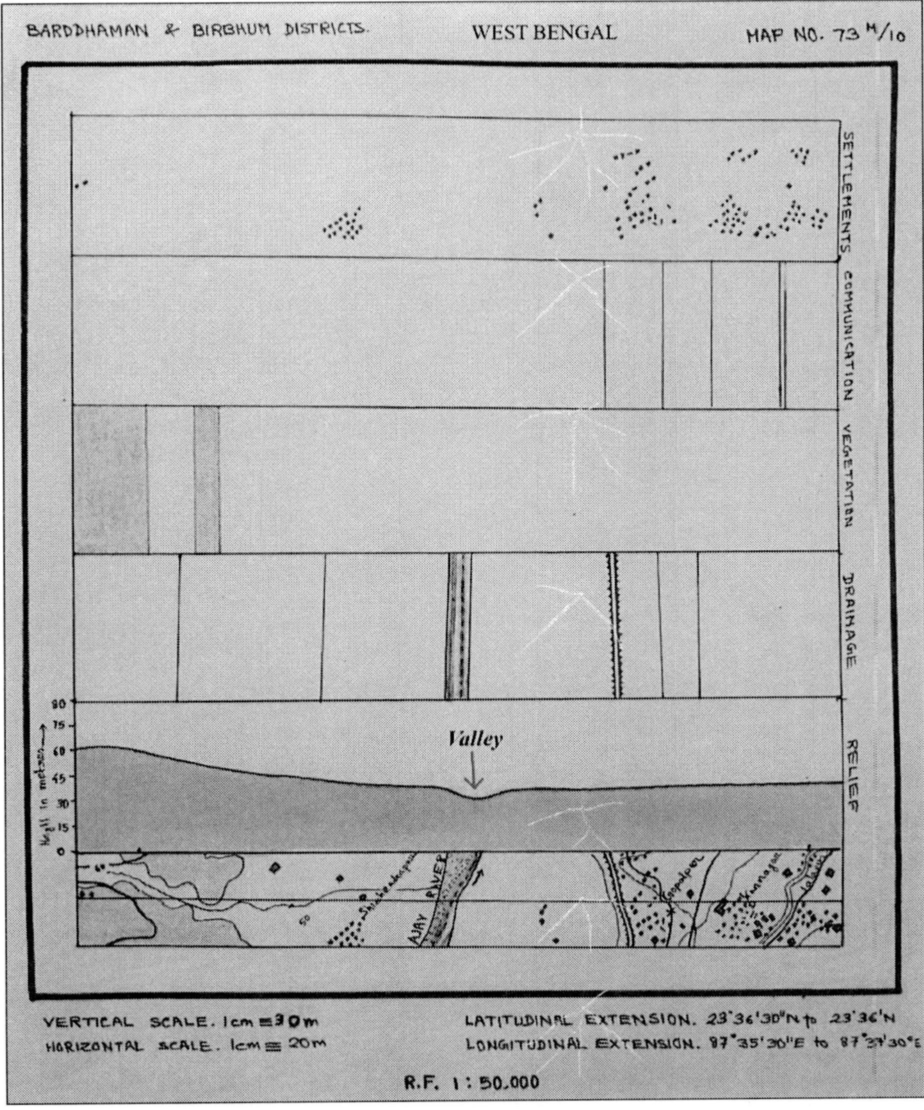

Figure 16: *Transect Chart depicting Interrelationship among Physical and Cultural Features of the Selected Area of the Topographical Map of 73 $^{M}/_{10}$*

Interpretation of Serial Profiles

Including Superimposed, Projected and Composite Profiles – A Specimen

Problem: To study the serial profiles of the area which has a latitudinal extent of 24° 20' 40" to 24° 25' N and longitudinal extent of 83° 50' to 83° 55' E.

Data base: The topographical sheet number is 63 P/5. The area lies between 24°15' to 24°30' N and 83°45'E to 84° E. The map is published in 1930. The magnetic variations

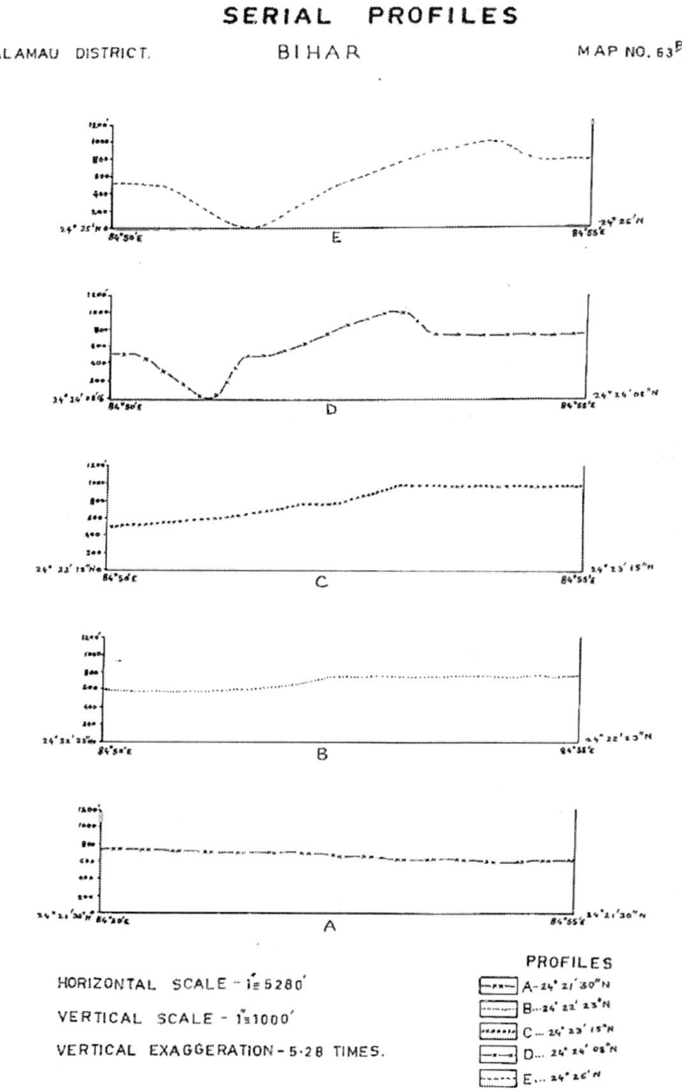

Figure 17.

from true North were about I' W in 1953. The contour interval is 50'. The region is situated in the Palamau district of Bihar. The scale of the map is 1 inch to 1 mile.

Objective: The main objectives of the drawing of serial profiles are:

(i) To show important geomorphic features of the area concerned

(ii) To compare and interpret the relief and

(iii) Finding out the probable controls of relief

Methodology

(i) Primarily, a typical block is taken for its depiction and detailed study.

(ii) Suitable cross-sections are chosen on the topographic map.

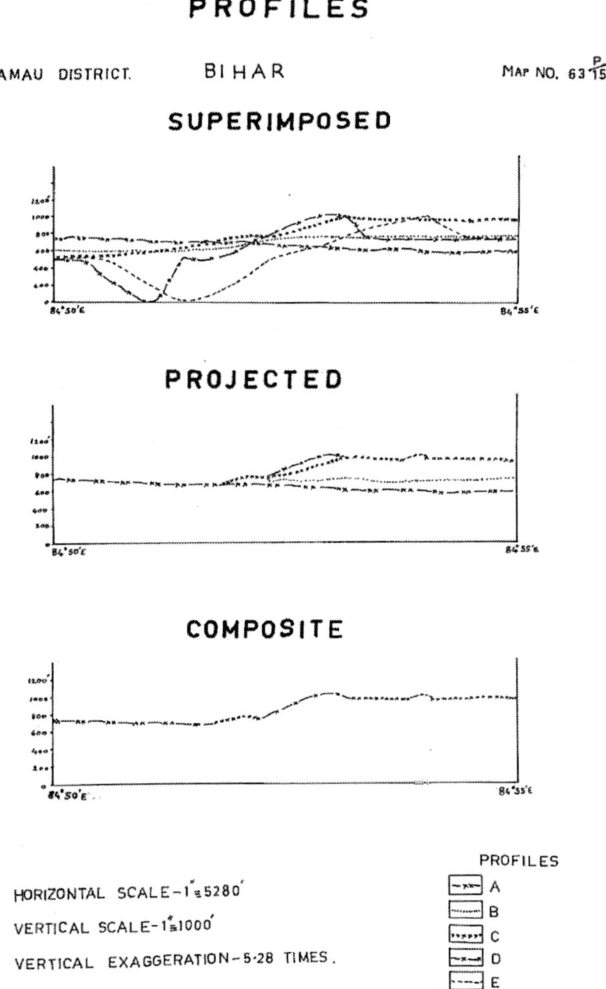

Figure 18.

(iii) Conversion of V.S. (1" ≡ 1000') and H.S. (1" ≡ 5280'). Thus, the Vertical Exaggeration (V.E.) becomes 5.28 times.
(iv) Five cross-sections have been selected latitudinally at definite intervals, having the same frame of longitudinal extension, and on the basis of these profiles – superimposed, projected and composite profiles are also drawn.

Interpretation: From these serial profiles ABCDE, it is evident that the topographic elevations of the plateau section are higher in the eastern part and lower in the western parts. This is due to the lithologic conditions and the geomorphic agent (fluvial action) curving on it. For better specificity names of fluvial channels and the sites may be relevantly mentioned.

In *superimposed* profiles, at the centre of the profiles D and E, it is more rugged ie. the evidence of the plateaux being dissected by the numerous non-perennial channels. While in the eastern portion, the profile exhibits a flat-topped resistant plateau section with one or two monadnocks. But in the western section, much of the relief of the plateau has been reduced due to the fluvial erosion of the main river present in the map and as a result, a relatively flat terrain has been formed whose elevation is about > 600'.

A *projected* profile from the south to the north has been drawn to compare the intervisibility. It exhibits only those features which are not obscured by higher intervening forms. It exhibits first an abrupt- increase of relief in the centre of the profile. On the western section, the land is relatively flat due to fluvial action. It gives a panoramic effect, with a foreground, middle ground, and distant skyline.

The *composite* profile shows only the ruggedness of the skyline. It is constructed to represent the surface as viewed in the horizontal plane of the summit levels from an infinite distance. So, from this profile, one can see that it is a highly dissected plateau area with isolated monadnocks, steep escarpments, and, gentle hill-slopes. But in the western section of the map, relief is comparatively low than in the eastern portion due to the lithologic conditions and the fluvial action on it.

11

Geological Map

A geological map shows the geographical distribution of different rock formations and strata at the ground surface. An appropriate Section Line exhibits the presence and thickness of all rock types in the area which led to the formation of the resultant topography. Its interpretation leads to an understanding of the lithology and the structure of the underlying rocks which help in the study of the evolution of landforms in a particular area.

In the following section, 2 Geological Maps have been discussed as specimens.

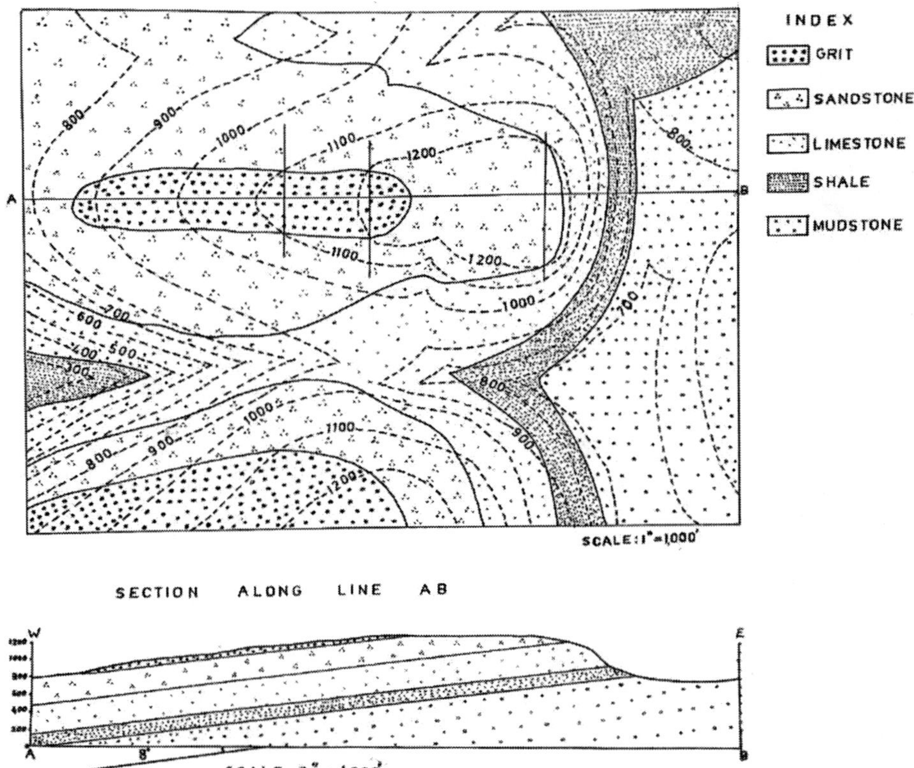

Geological Map No. 1

Introduction: The geological sections have been chosen along AB which runs west to east. It consists of a simple series and there is no unconformity.

(a) *Geological Descriptions: -*

Succession	*Vertical thickness*
Grit	100’
Sandstone	300’
Limestone	300’
Shale	100’
Mudstone	800’

(b) *Structure:* - The given section shows that it has a uniclinal structure which has a dip angle of about 8 degrees and the slope is from the east to the west.

(c) *Relation of rock group:* - It consists of sedimentary rocks like grit, sandstone, limestone, shale, and mudstone. The sequence of the beds is from the newest to the oldest which is grit, sandstone, limestone, shale, and mudstone.

(d) *Topography and its relation to structure:* - On the eastern side, there has been a valley due to the fluvial action of the soft rock like mudstone, whereas the limestone and sandstone being resistant have formed escarpments. Grit being softer has been exposed due to erosion.

(e) *Geological history:* - The beds have been laid down in order under marine conditions, and sub-ariel denudation has started forming the present topography.

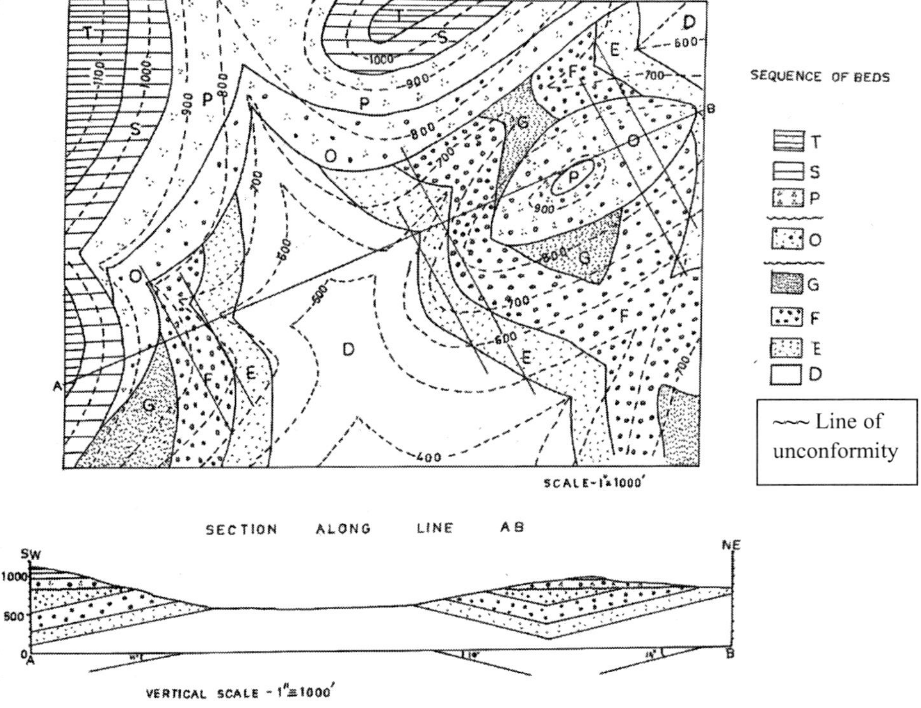

Geological Map No. 2

Introduction: The geological section has drawn from the southwest to the northeast along line AB. Due to folding; the dip of one bedding plane will vary from one limb of the fold to the next. The dip should, therefore, be determined for each limb of a fold.

(a) *Geological Descriptions: -*

Succession	*Vertical thickness*
T	undetermined
S	>50'
P	100'
~~	
O	
~~	
G	>500'
F	250'
E	200'
D	>700'

(b) *Structure:* - The beds show folded structure with an anticline in the central part of the section and a syncline on either side of it. The sedimentary beds represent a composite structure with the younger bed being horizontal and the older beds being folded. Within the older beds, there is one anticline in the mid-western part and one syncline in the eastern part. The western and eastern limbs of the anticlines dip at an angle of 11 degrees and 10 degrees respectively to the west and east respectively. The eastern limb of the syncline dips at an angle of 14 degrees towards the west.

(c) *Relation to rock groups:* - There are three series of rock beds that are separated by two lines of unconformities. T, S, and P comprise of the younger series; O is the middle rock layer and the rock beds of G, F, E, and D are the older series.

(d) *Topography in relation to structure:* - The region is characterized by ridge and valley topography. The ridge is located in the southwestern part and attains a height of over 900'. The valley has an average elevation of less than 600'. From the valley, the land rises gradually both to the east and west. In the western part of the land rises to a height of over 1100'. The valley occupies the anticlines while the ridge is located on the syncline. The anticlinal valley and the synclinal ridge represent an inversion of topography. Thus, it is evident that the topographical features have been structurally adjusted and the region of folded structure has attained the last stages in the process of the cycle of erosion. The younger beds occupy the highest parts in the southwestern parts, while the older beds have been successively exposed in the central and western parts. This has been due to differential erosion by the different geomorphic agents.

(e) *Geological history:* - The beds of the older series were deposited under the sea under shallow and marine conditions in the order from D to G. The beds were uplifted followed by folding. Sub-ariel erosion started, and the land has been peneplained and again submerged under the sea. The beds of the younger series were deposited over the beds of the older series. The whole area was again uplifted. Sub-ariel erosion started forming the present topography.

12

Survey Formats

Concept and Definition

Fieldwork or survey is an important pedagogical part of learning geography. Each and every region on this earth is unique in its physical and cultural elements. Therefore, to experience the traits of the natural region and to enhance the comprehension of the learners, field surveys are indispensable. To study the patterns and spatial distributions their interactions and relationships, survey works are essential. The studies can be done with any geographical issues like physical elements, settlement, economic activities, or contemporary societal issues.

The survey can be done at a local level or at a distant land. Surveys may differ in accordance to the fund outlay, purpose, grade, or capability of the learners and permissible time.

At the Primary, Secondary, and Senior Secondary stages of education, survey works encompass both primary and secondary sources.

The teacher has a vital role to guide and supervise the students or learners. The learners need to communicate well for their purpose and bear a positive gesture. All precautions and care should be taken so that no unwanted events occur and all data and information acquired are stored properly.

Formats of Questionnaire for Geographical Survey

Land-Use Study

Land is the most important resource. The primary, secondary and tertiary sectors of economy can be developed.

The land-use study is a particular method of study in which how plots of land are utilized is examined. In doing so, visit the plots of land personally and take care to collect information about the land from land-owners and local people with regard to its utilization. This can be done by visiting and observing the places and thus verifying the collected information.

After the collection of all information regarding the plots of land and their respective uses, need to be proceeded for drawing a map for it. This land use map will show how the land of an area has been used. It revealed the physical, social, economic, and other cultural aspects of the area.

Necessity of the Study of Land Use

Land-use study is an important part of Geography:

1. It helps to build up an important comprehensive geographical knowledge of the local environmental area.
2. A land-use study offers a clear idea about an area and its utilisation in both rural and urban perspectives.

3. It also builds up knowledge about the utility of land and the limitations of its development and some ideas about maps and land surveys.
4. The information table which comes out from the local land-use study often paves the way for further development of the plan of an area or at least an idea of it, often makes aware of the environment.

Procedure of Land-Use Study

First Phase: Collection of mauza maps or ward maps

The first step in this work is to collect a map of the area which is intended to study and know the land-use pattern. (a) If the region is a rural one, one has to collect a map of it where the mauza is located. For this, one has to collect a mauza-map according to J.L No. (Jurisdiction List Number) from the Collector's Office. (b) If the region is an urban area, one has to collect the ward map from the municipality or corporation in which it is located.

Second Phase: Drawing of the map, making it suitable for land-use study

The mauza-maps or ward-maps which are collected cannot be used readily as these are very small. These maps are usually in the scale of 40 cm = 1.6 km = 1600 cm or 1cm = 40 metre. As the scale of the map is too small to work, enlargement along its respective *Dag Number* (Plot No.) to make it usable for the study is necessary.

Third Phase: Selection of symbols or signs for the representation of the information collected

The selection of symbols or signs for the identification of the use and utilization of plot land is an important base work in this study. The land is used for various purposes like agriculture, housing, building roads, railways, water tanks, and for other services. Some lands remain as fallows. To mark various types of land accurately, selection of symbol or sign is necessary, such as, (i) Cultivated land-C; (ii) Housing land-H; (iii) Constructed building-B; (iv) Kutcha building (mud-built)-KB; (v) Pucca building (brick or stone built-PB; (vi) Road-R; (vii) Water bodies-W; (viii) Industrial estates or factories-I; (ix) School-S; (x) Hospital and health centre-HS; (xi) Post Office-PO; (xii) Religious Centre-RC; (xiii) Fallows-F, etc. For the perfect use of symbols, these selections may be grouped categorically in some graded classes as given below:

1. *Cultivated land (C):* Rice field-(R), Wheat field-C(U), Sugarcane field-C(S), Jute field-C(J), Vegetable field-C(V), Mango garden-C(M), Flower garden-C(F).
2. *Housing (H):* Kutcha (mud built) House-H (K), Pucca (brick built) House-H (P), Two-storied Pucca House-H (P2), Two-storied Kutcha House-H (K2), Three-storied Pucca House-H (P3), Three- storied Kutcha House-H (K3), School Building-H (S) etc.
3. *Road (R):* Unmetaled Road-R (U), Metaled Road-R (M), Brick built Road-R (B) etc.
4. *Water bodies (W):* Pond-W (P), Tube well-W (T), Canal-W (C), River-W (R) etc.
5. *Market-place, Business centre and Shop (M):* Market area-M (a) Shop-M (s), Factories-M (f), Cloth shop-(c).
6. *Religious Centre (RL):* Temple-RL (T), Mosque- RL (NI), Church-RL (C) etc.

Fourth Phase: Collection of information by visiting and observing the plots personally

In this phase, student should visit the place personally and inspect the plots and see how these are utilized and what purpose do they serve. Following the ward number or dag number, students will visit the plots and collect information which they will write in the survey note book and insert the symbols on the map.

Fifth Phase: Drawing of the map for the area surveyed
In this phase, students will draw a land-use map from the information so collected. In drawing the map, different shades, colours and symbols are to be employed to show the land-use pattern. An index should be furnished on one corner of the map, giving an explanation of colours, shades, and symbols.

[Directions and instructions about the use of colour: There are some different directions for the use of colour in the map, such as yellow or green for agricultural land, red for houses, buildings, and settlements, grey for fallows, blue for water bodies, rivers, canals, ponds and lakes. Again for showing different crops, such as rice, wheat, jute, coffee, sugarcane, oilseeds etc. different symbols may be used]

One or another corner of the map, name of the area, district, mauza or ward no. should be given clearly. North-Line direction should be shown at the one corner by an arrow.

After completion of the drawing usually on the bottom, the student should write his or her name of the school, class, section, roll number, and date.

[The land-use study of the rural and urban areas shows the following distinctive features:

a) In rural areas, most of the lands are utilized for agriculture; there are few roads and lines of communication as well as settlements and habitable areas.
b) In urban areas, most of the lands are occupied by habitable areas, houses, buildings, markets, shops, business centres, factories, etc. Open land and fields are hardly seen there.]

A Household Survey of a Locality

[Family Size/ Age-Sex Structure/ Educational Background/ Occupation]

There is a socio-economic survey along with the study of land-use of a region; the study on economic and social aspects is essential to have comprehensive knowledge about a region which is intended to study for a complete geographical account of a particular area.

Procedure of Study

1. *First Phase of Work: Preparation of questionnaire*

It is the first and foremost task for this survey. A series of questions relating to the social and economic conditions of the people of the region (which is going to be surveyed), is to be prepared. Questions may be set on the following items [A specimen of this questionnaire is given in the following.]

a) Name of the family, address, age, occupation, educational qualification, type of work, number of family members employed, monthly income etc.
b) Condition of the house (Kutcha-Mud-built, Pucca-Brick built- One or two storied), arrangement of drinking water, mother tongue, religion and, other matters if any.

In the case of rural studies, rural conditions, such as the total land area in possession, cultivated land, land used for housing, land used in different crop production, name of crops cultivated, the amount of agricultural income, amount of agricultural expanses, profit and loss etc. Considering the nature of place and economic conditions, a few more questions may be set, for an accurate survey, such as,

a) In the case of a shop or shopkeeper, a few more questions may be interrogated on the following: (i) Name of the shop and shop-keeper (ii) Name of the items which the shop-keeper deals in (iii) Number of buyers and from where they come (iv) Income, loss and profit etc.
b) In the case of factories and manufacturing centres, questions may be asked like (i) Name of the firm (ii) Producing items (iii) Raw materials used (iv) Volume of raw materials required (v) Source of these raw materials (vi) Number of workers employed (vii) Amount of investment (viii) Income, profit, loss etc.
c) In case of any problem or issue in the local or global community, the following questions may be arranged are like: (i) Name of the problem, (ii) Cause of the problem, (iii) Present sources of the problem, (iv) Arrangement to combat or handle the situation by local community or state level or central level government or NGOs, (v) Medical or financial assistance (if any) sources, (vi) Symptoms (acuteness or chronic) (vii) Mitigation or problem-solving techniques by different sections of the society, (viii) Any particular organisation involved, or their contribution, (ix) Mention the role of education in this field.

Surveyor should make comment on the findings of his or her survey and studies and should write everything and state whether the statements made by the local people are true or not. It may appear to the surveyor that the statements made by local people are not matching with me data or information collected by the surveyor. In this case, the surveyor's comment may help to evaluate the proper conditions.

After doing these, the surveyor should write his own name, name of the institution, class, section, roll number, data etc.

2. *Second Phase of Work: Door-to-door visits and collecting information from various sources*

In this case, the surveyor should proceed with the field study with his or her questionnaire and should visit door-to-door with his or her questionnaire and collect information from various sources. In doing so, he or she would make a friendly gesture and behave accordingly.

3. *Third Phase of Work: Setting up the information so collected in a graded pattern*

It is the most important phase of work. The surveyor should write down the collected information in a graded pattern and the population mentioning male or female, the standard of education mentioning the number of literate or illiterate persons and the incomes of the surveyed people.

4. *Fourth Phase of Work: Writing a description of the study*

In this phase, the surveyor should write the study report in the form of a description with the help of the findings in his or her study, mentioning the following information: (a) location of the area (village, mauza, P.S, district, J.L. No. etc.), (b) objective of the study, (c) feature of the socio-economic condition of the region, (d) general description relating to the mode of living of the inhabitants, environmental influences on them, nature of adaptation of the people to the geographical environment etc.

The study of land-use and socio-economic conditions presents some aspects of the comprehensive idea about the region. It helps in the study of Geography and enables one to realise the relationship of man to the local environment.

Suggested Topics for the Survey

1. Visits to concerned polluted sites, interviewing inhabitants, and preparation of reports of the environmental problems (causes, how affecting flora and fauna, and their solutions.
2. Visits to industries and studies on alternative strategies of environmental management.
3. To study the contribution of NGOs in improving a particular issue in a local area.

Questionnaire on Socio-Economic Condition

Date: ______	Serial No.: ______	
Name of the Surveyor: ______		
Name of the School: ______		
Class: ______	Section: ______	Roll No.: ______

1. Name of the head of the family:		13. Head of the family: Tiller ☐ Worker ☐
2. Age:		14. Land-holder: Yes ☐ No ☐
3. Number of the family members:		15. Qty. of land:
Male	Adults ☐ Minors ☐	Where:
Female	Adults ☐ Minors ☐	Description:
4. Mother-Tongue:		16. Name of crops cultivated:
5. Religion:		17. Qty. of production:
6. Schedule Caste or Tribe: Yes ☐ No ☐		18. Income from land:
7. Number of educated member(s):		19. Income of the others:
8. Highly educated:		a)
Doctor:		b)
Engineer:		20. Income from other sources:
Other:		a)
9. Birthplace of the head of the family:		b)
District:		21. Occupation of the other members:
State:		a)
10. Causes of migration:		b)
a) Job:		22. Unemployed, if any, in the family:
b) Education:		How many:
c) Other:		Female:
11. Monthly family income:		Male:
12. Occupational qualification of the head:		23. House (Kutcha/Pucca):
a) Service:		Single/Two Storied:
b) Place :		Owner/Tenant:
c) Business:		24. Roof made of:
d) Type of business:		(Thatched/Tally/Tin):

25. Is there any electricity/internet facility (as relevant):	32. Source of water:
26. Means of lighting:	Municipal water:
27. Mode of conveyance:	Tank:
Train/Bus:	Others:
Cycle/Car:	33. Distance of market or shop:
Foot/Any other:	34. Any other market nearby:
28. How do children go:	35. Inconveniences of that area, if any:
29. How far is their school:	a)
30. Classes studying:	b)
31. Highest degree attained in family:	c)

[*N.B.* Relevant questions may also be attached in accordance with the purpose of the study]

Remarks of the Surveyor:
Signature of the Surveyor:
Name of the School:
Class:
Section:
Roll No.:

____
Signature of the Geography Teacher
Date:

Rural Land-Use Survey

Date of Survey: **Time:**

Mauza: ________ J.L. No.: ________ Police Station: ______ District: ______
Location: ____________ Areal Extent: ____________
Relief: ____________ Soil: ____________
Drainage/River: ________ Canal: ________ Pond: ________
Total Inhabitants: ____________ Density of Persons per sq. km. ________
Male: ____________ Female: ____________
Literate: ____________ Illiterate: ____________
Number(s) of Plots/Plot No. (As Required): ____________

Identifying Household Plots:
Settlement: ________ Pucca: ________ Kutcha: ________
School: ________ Primary: ________ Madhyamik/School Final: ______
Post Office: ________ Health Centre: ________ Bank: ________
Clubs/ Organizations: ____ Panchayat Office: ______ Any Other Office: ________
Shops: ________ Fair/Market: ________
Temples: ________ Mosque: ________

Identifying Agricultural Plots:
Rice Cultivation: ______ Jute Cultivation: ______
Vegetable Cultivation: ____ Other Crops: ______

Identify Other Land-use:
Playground: ________ Pond: ________Garden Barren Land: ________
Use of Fertilizer: ______ Production of Crops: ____

Occupation of the Inhabitants:

Agriculture: ____________ Business: ______________Service Inherited/Family: ________
Handicrafts: ____________ Small Scale: ____________Animal Husbandry: ____________
Poultry: _______________ Transport Activities: _____

Remarks of the Surveyor:
Signature of the Surveyor:
Name of the School:
Class:
Section:
Roll No.:

____
Signature of the Geography Teacher
Date:

Urban Land-Use Survey

Date of Survey: **Time:**

Name of City: ______________ Police Station: ____________
Location: _________________ Name/No. of Ward: ________
Areal Extent: ______________ Population/No. of Inhabitants: _______________________
Literate: _________________ Illiterate: ________________
High-Income Family: _______ Middle-Income Family: ____ Low-Income Family: ______

Residence:

Temporary/ Unstable Huts: ___ Pucca House: _____________
Park: _____________________ Playground: ______________
Temples: _________________ Mosque: _________________ Church: _________________
Clubs/Organisations: ________ Library: ________________
Theatre Halls: ______________ Cinema Hall: _____________
Offices: ___________________ Shops/Shopping Mall: ______ Market: ________________
Large-Scale Industries: ______ Small-Scale Industries: _____ Cottage Industries: ________

Total Schools:

Primary Schools: ____________ Secondary Schools: ________ Colleges: _________________
Universities: _______________

Transportation Mode:

Metalled Roads/ National Highways: ____________________
Bus: ______________________ Trams: __________________ Minibus: ________________
Auto-Rickshaw: ____________ Rickshaws: _______________ Waterways: ______________
Private Vehicles: ____________ Cycle: ___________________ Motor-Cycle: _____________
Cars: _____________________ Metro: __________________ Rail-Station: _____________
Airport: __________________
Supply of Drinking Water: ___
Cemetery: _________________ Graveyard: _______________
Occupation of the Inhabitants: ________________________
Service: ___________________ Business: _________________ Transport Activities:_______
Inherited Occupation: _______ Handicrafts: _____________
Large/ Middle/ Small Scale Industries: ___________________

Remarks of the Surveyor:
Signature of the Surveyor:
Name of the School:
Class:
Section:
Roll No.:

____
Signature of the Geography Teacher
Date:

Survey of Agricultural Farm

Date of Survey: **Time:**

Name of the Teacher/Group Leader: ______
Total No. of the Students in Survey: ______
Name and Address of the Agricultural Plot Owner: ______
Area Extent of Land: ______
Soil Type and Characteristics: ______
Name of Crops during Survey and Type: ______
Sowing/Harvesting Time Period: ______
Cropping Techniques: ______
Type of Irrigation Applied: ______
Time of Applying Fertilizer (for Particular Crop): ______
Name of Fertilizers/ Insecticides: ______
Ripening Time of Particular Crop(s): ______
Estimated Production of Crop(s): ______
Production Cost Incurred by the Farmer/ Owner of the Plot: ______
Net Profit/Loss: ______
Other Producing States/Areas of the Particular Crop: ______
Comparison of the Crop in Relation to State/District Production: ______
Uses of the Crop: ______

[*N.B.* Relevant questions may also be attached in accordance with the purpose of the study]

Remarks of the Surveyor:
Name of the Student/Surveyor:
Class:
Section:
Roll No.:
Name and Address of the School:

______
Signature of the Geography Teacher
Date:

Survey on Exhibition / Market

Date of Survey: **Time:**

Name of the Teacher/Group Leader: ______
Total No. of the Students in Survey: ______
Name and Location: ______
Post Office: ______ District: ______
Occurrence of the Event (Time of the Year): ______
Duration of the Event: ______
Name of the Organiser: ______
Areal Extent of the Event: ______
Special Features/Points around the Event Ground: ______
Transport Linkage: ______
Rail Routes: ______ Road Routes: ______ Water Routes: ______
Name of the Participating Companies: ______
Products/Commodities Sold: ______
Special Incentives/Items Prevent: ______
No. of Seller: ______
No. of Customers (Buyers): ______

Advantages of the Event: ____________________
Defects of the Event: ____________________
Improvements of the Event: ____________________
Remarks (if any): ____________________

[*N.B.* Relevant questions may also be attached in accordance with the purpose of the study]

Remarks of the Surveyor:
Name of the Student/ Surveyor:
Class:
Section:
Roll No.:
Name and Address of the School:

____
Signature of the Geography Teacher
Date:

Survey of an Industrial Plant

Date of Survey: **Time:**

Name of the Teacher/Group Leader: ____________________
Total No. of the Students in Survey: ____________________
Name and Address of the Industry: ____________________
Type/Name of the Ownership: ____________________
Products (Processed/Finished): ____________________
Sources of Capital to the Plant: ____________________
Amount of Capital Involved: ____________________
Fuel/Power Supply: ____________________
Raw Materials Used: ____________________
Sources of Raw Material to the Plant: ____________________
No. of Labourers: ____________________
Cost of Production of Product(s): ____________________
Utility of the Finished Products: ____________________
Certified/Non-certified Products: ____________________
Marketing Network of the Products: ____________________
Financial Outlay of the Finished Products: ____________________
Net Profit/ Loss: ____________________

[*N.B.* Relevant questions may also be attached in accordance to the purpose of the study]

Remarks of the Surveyor:
Name of the Student/ Surveyor:
Class:
Section:
Roll No.:
Name and Address of the School:

____
Signature of the Geography Teacher
Date:

13

Importance of Studying with Relevant Typicals in Geography for Secondary Classes at Schools

Typicals are items or things that have distinctive qualities/traits and represent a particular group or category/class. Their signs, symbols, pictures, or diagrammatical representations or affixing of real-life objects (in very miniature forms) may be presented to depict the 'typical' nature. As these items/objects have the ability to exhibit similar/related essential characteristics pertaining to geographical concerns, they may help the learners of different grades to familiarize themselves and get sensitization about the various environmental issues. These *activity-based tasks* are related to collecting information or pictures/typicals, appropriate to the concerned geographical topic of study from different sources, generate curiosity and facilitate inquiry-based learning, add interest and also enable the understanding of the geographical concepts.

It is worth mentioning that these typicals (specimens) can be prepared appropriately and arranged by the subject expertise for an inclusive classroom in order to enhance and enrich the learning process.

Suggested Typicals in Geography:

1. Common fibres (*given as examples*) - Given as specimen
2. Digital assets at a glance (*given as examples*) - Given as specimen
3. Soil types & texture (*related to climatic and vegetation influences*)
4. Shapes and colours of plant leaves (*related to climate*)
5. Identification of minerals or rocks (*real-life miniature forms*)
6. Costumes and festivals (related to climate - *statewise*)
7. Flow, fund, and alternate sources of resources (energy)
8. House types (related to climate)
9. Items of Disaster Management Kits, Road Safety symbols, First-aid kit, COVID-19 Medical Kit
10. Species related to endangered/extinct/aerial/arboreal/deep sea
11. Modern Surveying types of equipment
12. Management of bio-degradable & non- bio-degradable wastes.

Different Disciplinary Perspectives on the Study of a Consumer Product – (*Activity Based On Thinking Skills*)

Problem: To trace any consumer product from its raw form and to present the study from different disciplinary perspectives.

In this case, the major disciplines taken for study are Geography, Biology, Physical Science, Economics, History, Politics, Environmental Education, Mathematics/ Statistics, and Art, and the consumer product selected for discussion is Petroleum – The Liquid Gold

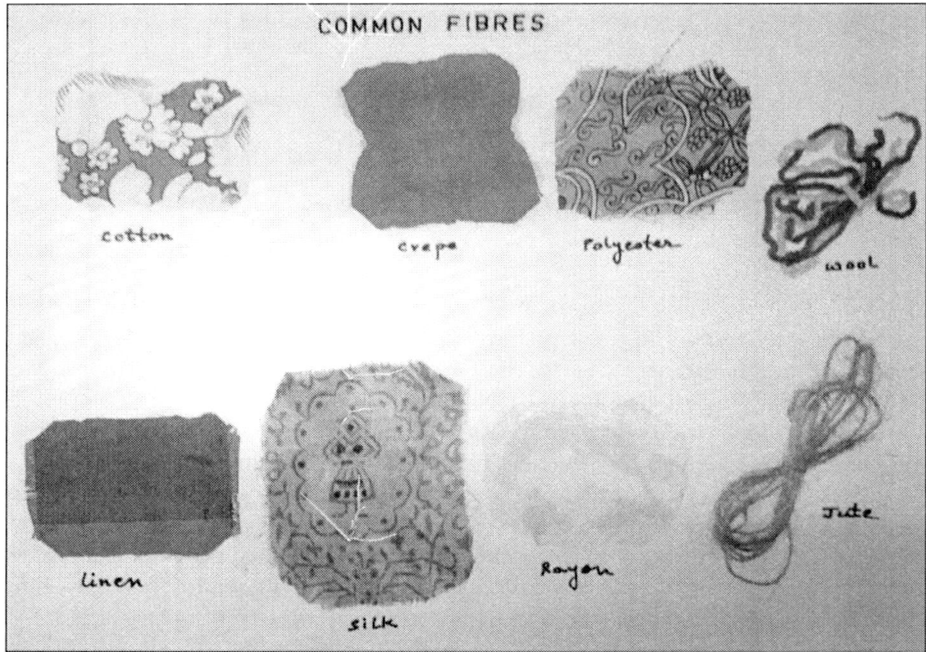

Figure 1: *Common Fibres*

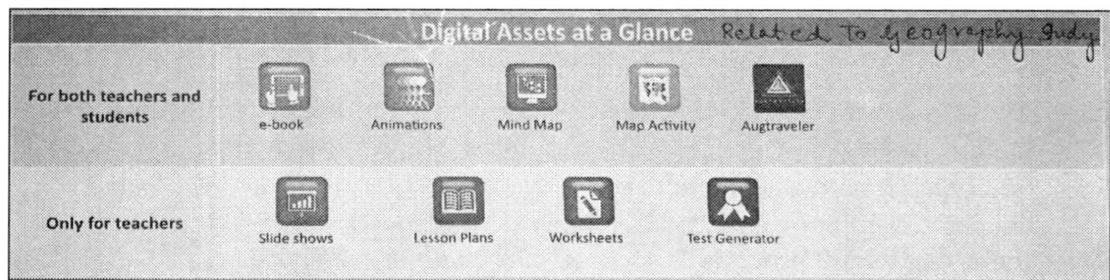

Figure 2: *Digital Assets at A Glance*

Geography

1. Properties and Nature
2. Oil prospecting and drilling
3. Oil transportation and storage
4. Distribution/ of oil belts with/without maps.
5. Problems and presence of oil/petroleum industries
6. Uses of oil

Biology

1. Origin/occurrence/formation of oil
2. Discoveries of new oil fields

Physical Science (Physics and Chemistry)

1. Oil refining process
2. Research and development for commercial production of by-products
3. Scientific and infrastructural facilities for the discovery of new oil fields

Economics
1. Factors of the location of oil refineries
2. Location of petrochemical industries
3. Economic consequences of the large oil fields in sparsely populated areas
4. Demand and supply chain of old and new oil fields
5. Concept of optimum utilization and market strategies
6. Fund allocation for flow resources/ non-renewable sources of energy
7. Government intervention in the context of the energy crisis

History
1. History of oil extraction from the earliest time till modem technological facilities
2. Increasing role of petroleum as a domestic and industrial fuel with the declining importance of coal
3. Formation and organization of OPEC

Politics
1. OPEC and energy crisis
2. Future of oil industries
3. Oil prices and investments in global markets
4. Exporting and importing countries—issues related to crude/refined oil

Environmental Education
1. Oil spill/oil pollution- causes, description, mitigation, and future preventive measures
2. Legal measures- Acts/Rules for environmental safety of oil transaction
3. Environmental protection in respect of the discovery of new fields

Mathematics/Statistics
1. Production/Trend of demand and supply of oil/petroleum products in the world
2. Consumption pattern and prices (comparison) with other fuels
3. Cost-Benefit analysis (comparison with alternate fuels)

Art
1. Drawing of occurrence of oil in rock layers (permeable and non-permeable)
2. Depiction with appropriate symbols of distribution of oil fields
3. Graphical representations of the export/ import or price trends of oil in different countries
4. Timelines showing the production of by-products of petroleum with time

Great Geographers

Importance of Study of Great Geographers' Contribution

It is noteworthy to study the contributions of the Great Geographers who have given varied dimensional thought and have rediscovered the subject from different perspectives. This scientific intervention provides a blueprint for the future of the discipline, recommending how to strengthen its thought-provoking domains so as to study natural and cultural relationships. Through the curriculum of Geography teaching, the "locality-sense" can be gradually developed to the horizon of global thinking. To open-up innovative avenues of research and build sub-disciplinary ideas and adoption of contemporary learning tactics an ideal geographer/ teacher can lead a learner virtually all over the globe through streamlined appropriate strategies.

ERATOSTHENES
c. 276 - c. 195 BC

He invented the system of Latitude and Longitude; calculated the circumference of the earth and tilt of the earth's axis; and was the first to use the word 'Geography'

CARL O' SAUER
1889-1975

He developed the idea of 'Cultural Landscapes' in his book, THE MORPHOLOGY OF LANDSCAPE (1925)

CARL RITTER
1779-1859

He developed the idea of 'Unity in Diversity', and declared geography as 'ERDKUNDE' i.e. an earth science

HALFORD MACKINDER
1861-1947

He was the founding father of 'Modern Geo-politics' and 'Geo-strategy'. He formulated the concept of 'Geographical pivot of History' or 'Heartland Theory' (1904)

ALEXANDER VON HUMBOLDT
1769-1859

He was a great traveller and wrote KOSMOS (1845) and laid the foundation of 'Bio-geography'

RICHARD HARTSHORNE
1899-1992

An ardent advocate of regional approach in geography, he wrote the famous book PERSPECTIVE ON THE NATURE OF GEOGRAPHY (1959)

DAVID HARVEY
b. 1935
Advocate of the 'Right to the City', he is the world's most cited geographer and social thinker. His famous books include EXPLANATION IN GEOGRAPHY (1969) and SOCIAL JUSTICE AND THE CITY (1973)

ELLSWORTH HUNTINGTON
1876-1907
Postulated the idea of 'Climatic Determinism' in his famous book CIVILIZATION AND CLIMATE (1915)

FRIEDRICH RATZEL
1844 - 1904
He propounded the concept of 'Social Darwinism' and wrote two seminal books, ANTHROPOGEOGRAPHIE (1882-91) and POLITISCHE GEOGRAPHIE (1897)

SIVAPRASAD CHATTERJEE
1903-1989
He founded the Department of Geography, Calcutta University (1941), the National Atlas & Thematic Mapping Organization (1956) and was awarded with PADAMBHUSAN

VIDAL DE LA BLACHE
1845-1918
He was known as the 'Founder of Human Geography', as he pleaded for the notion of 'Possibilism'

WILLIAM MORRIS DAVIS
1850-1934
He postulated the concept of 'Cycle of Erosion' in development of landforms. He was the most influential geomorphologist of his time, and he founded the Association of American Geographers (1904)

14

Statistical Representations

Meaning of Statistics

Statistics is a very familiar word nowadays. The word 'statistics' is associated with the term indicating quantitative information. The term Statistics was coined by the German scholar Gottfried Achenwall around the middle of the eighteenth Century. In the modern context, the word includes all branches of human activity concerned with numerical data. Statistics is used as a plural noun and also as a singular noun. It has two different meanings, in the plural sense, the word implies a set of numerical figures usually obtained by measurement or counting connected with any field of human experiences. In a singular sense, Statistics refers to the subject of scientific activity which deals with the theories and methods of collection, presentation, analysis, and interpretation of numerical data.

Functions of Statistics

- *To simplify complex data*- The fundamental function of the statistical method is to reveal the significance of complex data by means of various diagrams, different averages, etc.
- *To compare*- Statistics facilitates the comparison of data both chronologically and geographically.
- *Formulation and testing of hypothesis*- Statistical methods have their use in formulating and examining hypotheses.
- *Prediction*- Statistical techniques are helpful in forecasting future trends in different fields of activity and thus in relation to industry, business, developmental plans, and policies, they can be formulated well in advance of their time of execution.
- *Testing the laws of other Sciences*- Statistics is very useful for examining the facts of all sciences whether Natural or Social Sciences.
- *To conduct a survey*- Statistical methods are used in making the efficient design of experiments and surveys and to interpret the relevant information for better social development.
- *To help the government*- The important principles of Statistics have become necessary for the administration of a country.

Importance of Statistics

Statistical tools and techniques are found to be applicable in different fields of Business, Commerce, and Management for academic pursuits and also in scientific interventions.

Limitations of Statistics

It is applicable only to quantitative data and mainly related to the aggregate of facts and is not concerned with individual items. Secondly, statistical laws are true only on average and they may not hold in a particular situation. Thirdly, in a particular case, statistical methods, if not applied in the proper perspective of the collected data, may lead to false conclusions. Fourthly, though they are useful for prediction and are affected by a multiplicity of causes, exactness, cannot be guaranteed, and no estimates or forecasts for statistical decisions can be taken to be exact. They are all subject to a certain amount of inaccuracy whose limits may be stated in terms of probability. Lastly, statistical data must be uniform in the sense and need to be handled with care by experts or by learners and care should be taken so that it is not misused in any research pursuits.

Data

*Data m*eans a body of information collected in numerical form and that need to be arranged in a tabular form called *Data Matrix*. The unclassified data from which frequency distribution may be done may have an unsystematic arrangement of R*aw Data* and irrespective of any order. An *Array* is defined by an arrangement of data in which all items are placed sequentially in order of magnitude. This array helps one to know at a glance, the range of values, the nature of concentration, and the degree of continuity of the data and their distribution pattern.

Collection of Data

The collection of data is the primary task at the outset of any statistical activity which can be collected through scheduled questionnaires and pilot studies.

Primary Data and Secondary Data

It may be classified as Primary and Secondary, depending on the sources of the data.

1. Primary – The statistical data which are gathered directly from the field of investigation for the desired purpose is called primary data and these are obviously original in nature and have the reliability of data.
2. Secondary – Data collected by some agency and compiled from that source by the enquirer for his/her use, are called secondary data. These data are original and cheaper to collect and involve less manpower and time.

Variables

The items about which information has been collected are usually referred to as individuals. In the data matrix, each column gives the value of a specific property or characteristic that varies from one individual to another. These are called variables. The variables that can take on all sorts of fractional values are called continuous variables, for example, height, length, etc. In discrete variables, the values are all simple whole numbers so that no score can fall in the gaps between each pair of successive whole numbers.

Variables may also be classified as discrete or discontinuous and continuous.

1. Discrete or Discontinuous variable - A quantitative character that can take certain isolated values only in its range of variation is called a discrete variable. Example: Size of families of some locality, number of students in different colleges.
2. Continuous variable - A variable that can assume any value within its range of variation is termed a continuous variable. Example: Weights of individuals or the marks of candidates.

In the case of discrete variables frequency distribution can be represented by perpendiculars (or columns) having heights equal to the frequencies of the variable values which are drawn at the corresponding points (indicating variables values on the horizontal axis). The diagram so formed is called a column diagram or a frequency diagram.

Class

In organizing and summarizing that statistical data, a frequency distribution is normally constructed. In this construction process, the whole range of the data is divided into some groups defined by an interval, and the number of observations falling in each group is stated precisely. The groups are called *class intervals* or simply *classes* and the observations are called *frequencies*. The classes are defined by limits or boundaries on two ends. When one end of the class is not specified, it is called an *open-end class*. Normally in distribution, all classes are of the same size but in some cases, classes of *unequal* size may also be constructed, particularly for *highly dispersed* data. The number of observations included in a class is termed as absolute frequency or *frequency of the class*. The *class limits* are the two end-points of a class interval used for tally marking the given values. The *class boundaries* are the two real endpoints of a class interval. The upper-class boundary coincides with the lower-class boundary of the next class. The class boundaries are used for forming the frequency distribution of a continuous variable.

These are defined as the most extreme values that would ever be included in a class. In fact, the class boundaries are the real class limits. The lower extreme value is called the lower class boundary while the upper extreme value is called the upper-class boundary. For all practical purposes of graphical representation of data, class boundaries are used. The working formula is:

$$\text{Lower Class Boundary} = [\text{Upper Class Boundary} - (d/2)] \text{ and}$$

$$\text{Upper Class Boundary} = [\text{Upper Class Boundary} - (d/2)]$$

where,
d = Difference between the two successive class limits.

Class Width: The width or size of a class is the difference between the lower and the upper class. It is expressed as,

$$\text{Class Width (i)} = (\text{Upper-Class Boundary} - \text{Lower Class Boundary})$$

Class Mark: The Class Mark or mid-value of a class refers to the value that lies exactly in the middle of a class. It is used as the representative value of a class for the purpose of calculation of descriptive measures, The common formula is:

$$\text{Class Mark (x)} = \tfrac{1}{2}\,[\text{Lower Class Limit} + \text{Upper Class Limit}]$$

or,

$$= \tfrac{1}{2}\,[\text{Lower Class Boundary} + \text{Upper Class Boundary}]$$

Class Frequency: The number of observations contained in a class is known as its class frequency. The sum of all class frequencies in a distribution is called total frequency. Classes with zero frequencies are empty classes. Therefore,

$$\text{Total Frequency, (N)} = \sum_{i=1}^{n} fi$$

Where,
n = Number of Class and
f_i = Frequency of the ith class

Frequency Distribution

Frequency distribution is a statistical table that shows the values of the variables and the corresponding number of observations or frequencies. Frequency distribution may be of two types – *simple and grouped.* In a *simple frequency distribution,* values of the variables are arranged individually in order of magnitude. In a *grouped frequency distribution,* values of the variables are arranged in groups in order of magnitude. Frequency distribution is generally used to:

i. Condense and summarize a huge body of data
ii. Visualize the range, average, spread and shape of the distribution and
iii. Approximate the nature of probability distribution in the population.

Relative Frequency: Relative frequency is simply the class frequency (f_i) expressed as the proportion of the total frequency (N) of a distribution. The sum of all relative frequencies in distribution is equal to unity. In expression form:

$$\text{Relative Frequency, R}f_i = \left(\frac{fi}{N}\right)$$

and,

$$\sum_{i=1}^{n} R.fi = 1$$

where, n = Number of Class

Frequency Density: The Frequency Density of a class is defined as its frequency per unit width. It indicated the concentration of frequency in a class. The common expression is:

$$Frequency\ Density\ (fi) = \frac{frequency\ (f)}{class\, width\ (i)}$$

Frequency distribution may also be represented in terms of *relative frequencies or cumulative frequencies.* The *cumulative frequencies* are found by successively adding the class frequencies, where the additions begin from the top (that is the lowest class) or bottom (that is the highest class). Depending on – *less-than type or more-than type cumulative frequencies* are obtained.

Histogram

It is an appropriate diagram for representing the frequency distribution of a continuous variable. It consists of rectangles, each of which has breadth equal or proportional to the size of the concerned class intervals and height equal or proportional to the corresponding frequency. In a histogram, consecutive rectangles have a common side and for this, the class intervals are made overlapping in all cases.

A histogram displays the frequencies of different class intervals. As such it becomes easier to locate the class intervals having the greatest frequency or the least frequency and also helps to find the *mode* of distribution.

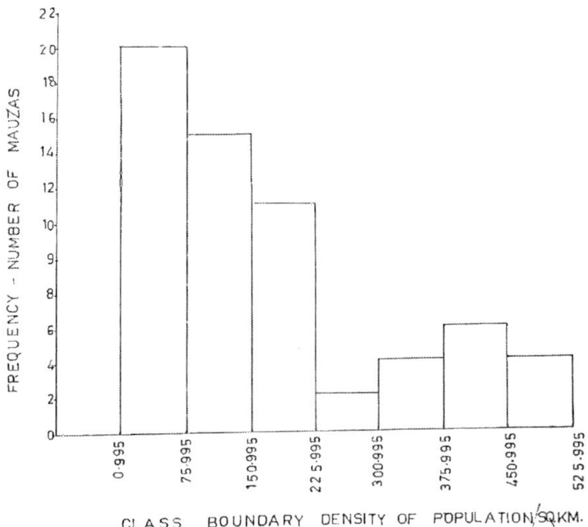

Histogram Showing Distribution of Mauzawise Population Density-P.S. Kalimpong Dist. Darjeeling, West Bengal - 1981

Frequency Polygon

It is a suitable diagrammatic mode of exhibiting the frequency of a distribution of a discrete variable. The variable values are located on the horizontal axis and the frequencies on the vertical axis in a rectangular system of coordinates. The available data are plotted as points on the plane where the abscissa and the ordinate for a point respectively indicate the variable values and the corresponding frequency.

Finally, when the points are successively joined by line segments the frequency polygon is obtained.

Therefore a *frequency polygon* is a polygon whose vertices are at the middle points of the tops of rectangles forming the histogram of the frequency distribution. These middle points correspond to the class marks of the corresponding class intervals.

There are two methods of constructing a frequency polygon for classified data:

(a) With the help of a histogram
(b) With the help of class marks

Interpretation - From the diagram, it is seen that the class boundary, 0.995 – 75.995 has the highest frequency of 22, whereas the lowest frequency is 2 lying within the class boundary of 225.995 – 300.995. The frequency polygon curve is double-humped, the first hump lies within the first-class interval, and the second hump in the second-last-class interval.

Ogive (Cumulative – frequency Curve)

Grouped data are also reprinted by a curve called *ogive or cumulative frequency curve.* In the case of less- than type cumulative frequencies, they are plotted against the upper-class boundaries as different points which are joined successively by line segments to

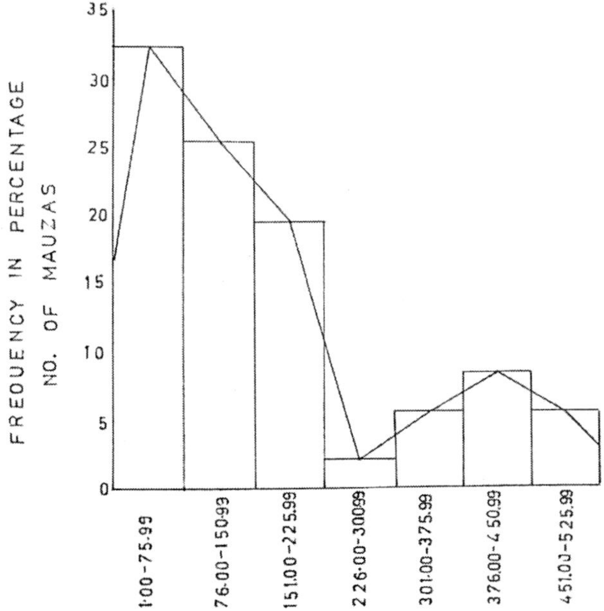

Histogram & Frequency Polygon Showing Distribution of Mauzawise Population Density-P.S. Kalimpong Dist. Darjeeling, West Bengal - 1981

get less-than-type ogive. Again, for a more- than type ogive cumulative frequencies of more-than-type correspond to lower class boundaries. Obviously, the cumulative frequency of the less-than type is zero for the lower boundary for the lowest class and it and the cumulative frequency of greater than type /more-than type are zero for the upper boundary of the highest class which has to be included in the drawing diagram.

An ogive for a frequency distribution displays cumulative frequencies of different class intervals and helps to find the median, the upper and lower quartiles of a distribution.

Shape of the Frequency Curve

The shape of a frequency curve is an important characteristic of a distribution. It is normally measured from two aspects – symmetry of the frequency curve (*skewness*) and peakedness of the frequency curve (*kurtosis*).

Skewness – the term skewness is used to denote the extent of asymmetry in the distribution any asymmetric distribution is a skewed distribution basically it measures the extent by which the bulk of the values in distribution is concentrated to one side or the other of the mean. A perfectly symmetrical normal distribution has, therefore, zero skewness. In any distribution, if Mean>Median>Mode, it is positively skewed whereas if Mean<Median<Mode, the distribution is negatively skewed. It has been shown in the diagram below.

Kurtosis – Kurtosis measures the extent to which values are concentrated in one part of the frequency distribution. Hence it refers to the degree of peakedness of the frequency curve. The two distributions may have the same average, spread and symmetry but they can represent different degrees of peakedness. For a normal distribution that is neither very peaked nor flat-topped, kurtosis is 3. This is called a *meso-kurtic* distribution. When kurtosis is less than 3 the frequency curve is flat-topped and the distribution is called *platy-kurtic*. When the frequency curve is highly peaked and the value of the kurtosis is more than 3, the distribution is called *lepto-kurtic*.

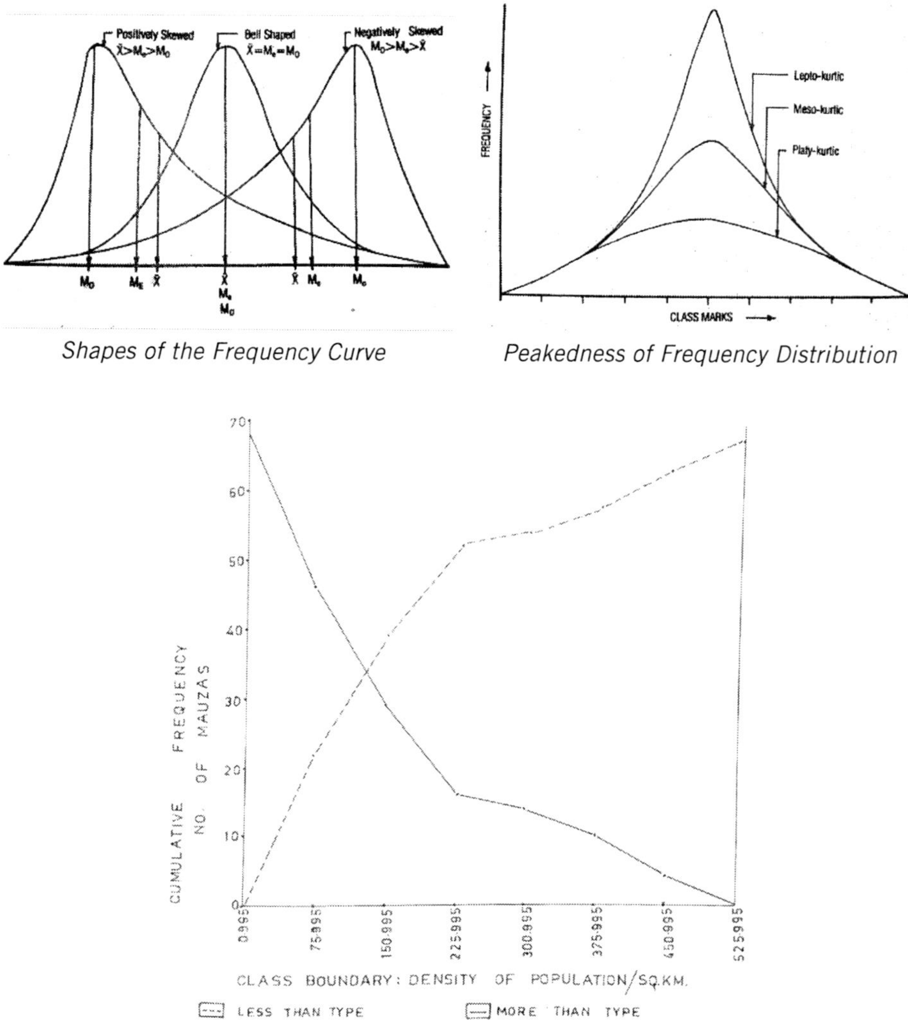

Shapes of the Frequency Curve

Peakedness of Frequency Distribution

Ogives Cumulative Frequency Curves Showing Distribution of Mauzawise Population Density-P.S. Kalimpong, Dist. Darjeeling, West Bengal - 1981

Data for Histogram, Frequency Polygon and Ogives Cumulative Frequency Curves of Mauzawise Population Density – P.S. Kalimpong, Dist. Darjeeling – 1981

Class Interval Pop/Sq. Km	*Class Boundary*	*Frequency*	*Cumulative Frequency (less than)*	*Cumulative Frequency (more than)*	*% of Frequency*	% C_f (L)*	% C_f (M)*
1 – 75.99	0.995 – 75.995	22	22	68	33.25	32.35	100.00
76 – 150.99	75.995 – 150.995	17	39	46	25.00	57.35	67.65
151 – 225.99	150.995 – 225.995	13	52	29	19.13	76.48	42.65
226 – 300.99	225.995 – 300.995	2	54	16	2.94	79.42	23.52
301 – 375.99	300.995 – 375.995	4	58	14	5.88	85.30	20.58
376 – 450.99	375.995 – 450.995	6	64	10	8.22	94.12	14.70
451 – 525.99	450.995 – 525.995	4	68	4	5.88	100.00	5.88

*L – Less than type, *M – More than type

Interpretation – Both the cumulative frequency curves (less-than and more-than) show a gradual stopping tendency. But, near the value of 25.995, the sloping nature of the curve changes, i.e. from a gentle-slope to a more gradual sloping curve.

Measures of Central Tendency

In order to make a comparative study of Frequency Distributions, arithmetic number are needed. Depending on the nature of the distribution, such arithmetic numbers are calculated in different ways. Such number are called arithmetic measures of data. These numbers indicate how the data are distributed around the central part of the distribution. So, these arithmetic numbers are known as *measures of central tendency* which are as follows: (1) Mean or arithmetic mean (2) Median and quartiles and (3) Mode.

Mean

A. Mean of Ungrouped (Raw) Data

The mean (or arithmetic mean) of n observations (variates) $x_1, x_2, x_3, \ldots, x_n$ is given by

Symbolically, $A = \frac{\sum xi}{n}$. In words, mean = $\frac{sum\ of\ the\ variates}{total\ number\ of\ variates}$.

Note $\Sigma x_i = nA$, *i.e.*, sum of the variates = mean × number of variates

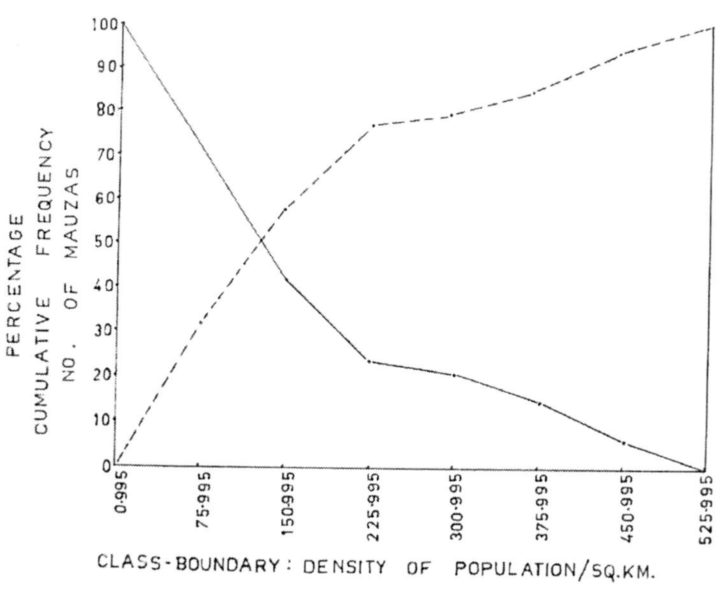

Ogives Relative Cumulative Frequency Curves Showing Distribution of Mauzawise Population Density-P.S. Kalimpong, Dist. Darjeeling, West Bengal - 1981

B. Mean of Grouped (Arrayed) Data

If the values of the variables (called the observations or variates) be $x_1, x_2, x_3, \ldots, x_n$ and their corresponding frequencies are $f_1, f_2, f_3, \ldots\ldots, f_n$ then the mean of the data is given by

$$\text{Symbolically, } A = \frac{\sum xifi}{\sum fi}$$

In words (by direct method),

$$\text{mean} = \frac{\textit{sum of the products of the variates and their corresponding frequencies}}{\textit{total frequency}}$$

Median

Like mean median is another measure of central tendency of a distribution. If the observations (variates) are arranged in ascending or descending order then the variate in the middle is called the *median* of the distribution. Thus the median divides the ordered observation in two equal halves.

i) If *n is odd* then the $\frac{n+1}{2}$*th* variate is the median

ii) If *n is even* then the mean of the $\frac{n}{2}$*th* and $\left(\frac{n}{2}+1\right)$*th* variates in the median, i.e.,

$$\text{median} = \frac{\frac{n}{2}th\,variate + \left(\frac{n}{2}+1\right)th\,variate}{2}$$

$$\text{median} = L + \frac{\frac{N}{2} - F}{f} \times i$$

where,

L = Lower boundary of the median class

f = Frequency of the media class

i = Interval of the median class

N = Total frequency

F = Cumulative frequency upto the median class

Quartiles

The three variates which divides the data of a distribution in four equal parts (quarters) are called quartiles. The median is the second quartile.

If the data are arranged in ascending or descending order then the variate lying at the middle between the lowest variate and the median is the called the *lower quartile* (or the *first quartile*), and it is denoted by Q_1.

The variates of a data are real numbers (usually integers) so they are scattered over a part of the number line. A researcher would like to know the nature of the scattering of the variates. The arithmetic numbers associates with distributions to show the nature of scattering are known as *measures of dispersion*:

(a) Range (b) Interquartile Range

(a) Range: The difference of the greatest variate and the smallest variate in a distribution is called the range of the distribution.

(b) Interquartile Range: The interquartile range of a distribution is $Q_3 - Q_1$, where Q_1 = lower quartile and Q_3 = upper quartile.

$\frac{1}{2}(Q_3 - Q_1)$ is known as *semi-interquartile range.*

Median of Classified Data

The median of a distribution in a continuous classified form can be calculated either by formula or by the ogive of the distribution.

Median class

If in a continuous distribution the total frequency be N then the class interval whose cumulative frequency is just greater than $\frac{N}{2}$ (or equal $\frac{N}{2}$) is called the *median class.* In other words, median class is the class interval in which the median lies.

Percentiles

Percentiles are the points which divide the set of observations into one hundred equal parts. These points are denoted by P1,P2,P3,...,P_{99}, and are called the first, second, third... ninety ninth percentile. The percentiles are calculated for a very large number of observations like workers in factories and the populations in provinces or countries.

Deciles

Deciles are similar to Percentiles (sounds like decimal and percentile together), as they split the data into 10% groups: The 1st decile is the 10th percentile (the value that divides the data so 10% is below it), the 2nd decile is the 20th percentile (the value that divides the data so 20% is below it).

A decile rank arranges the data in order from lowest to highest and is done on a scale of one to ten where each successive number corresponds to an increase of 10 percentage points.

Mode

Mode (or modal value) is another measure of central tendency. The *mode* of a distribution is the value represented by the point about which the variates tend to be most heavily concentrated. In general, the mode is the variate which occurs more frequently, i.e., the variate having the highest frequency. Mode is usually denoted by M_0.

Mode of Classified Data

The mode of a distribution in a continuous classified form can be calculated either by formula or by the histogram of the distribution.

Modal Class

The class interval having the maximum frequency is called the *modal class* and its class mark is called the *crude mode.*

$$\text{Mode} = L + \frac{(f_m - f_p)}{(f_m - f_p) + (f_m - f_n)} \times \text{i}$$

where,

f_m = Modal class frequency

f_p = Frequency of the class just preceding the modal class

f_n = Frequency of the class just next or following the modal class

L = Lower boundary of the modal class
i = Modal class interval

Comparison between Mean Median Mode

The mean and the mode are rigidly defined. The median is also rigidly defined, except when there is an even number of observations. In finding each of these measures, all the observations are taken into consideration. However, only the mean is directly based on all the values and its value changes even if a single observation is altered. On the other hand, median and mode may remain unchanged even after the alteration of several observations. The significance of all the measures is quite easily comprehensible. In general, the labour involved in the computation of all the three measures is almost the same. But in practice, the exact determination of mode of a continuous variable is impossible, because practically an ideal distribution (or the frequency curve) is never obtained. The mean has several properties and by virtue of them it can be readily manipulated in varied situations. But the median and mode does not possess such desirable properties. Among the three measures, the mean is generally found to be least affected by sampling fluctuations. However, in this respect, the median or mode may be better than mean in some specific situations.

Thus it is evident that, in general, mean is the best measure of central tendency. But there are situations where mean cannot be used or should not be used. In the case of a grouped frequency distribution, if one or both of the terminal classes are open, mean is indeterminate. Again, if there be a few extreme values markedly different from majority of the values, mean should not be used as an average. In such situations, median or mode would be the appropriate measure.

Quartile Deviation

The Quartile Deviation is defined as the semi-interquartile range. It is dependent on the upper and lower quartile values and does not concern the variability of the largest 25% or the smallest 25% of the observations. However, it takes into account 50% of the observations lying between the first and third quartiles.

$$\text{Quartile Deviation (QD)} = \frac{Q_3 - Q_1}{2}$$

Standard Deviation

The variance is not normally used as a descriptive measure, instead the square root of the various is taken. It is called standard deviation (σ) which is defined as the *root-mean-squared-deviation from the mean*. It is superior to variance in that variance may become a large number, even greater than the individual values but in the standard deviation it is square-rooted so it becomes more comprehensive and manageable.

$$\text{Standard Deviation } (\sigma) = \sqrt{\sum_{i=1}^{n} \frac{(x_i - \overline{x})^2}{N}}$$

Scatter Diagram

The simplest way of a diagrammatic representation of bi-variant data is by using a scatter diagram or dot diagram taking two perpendicular axes of coordinates, one for X

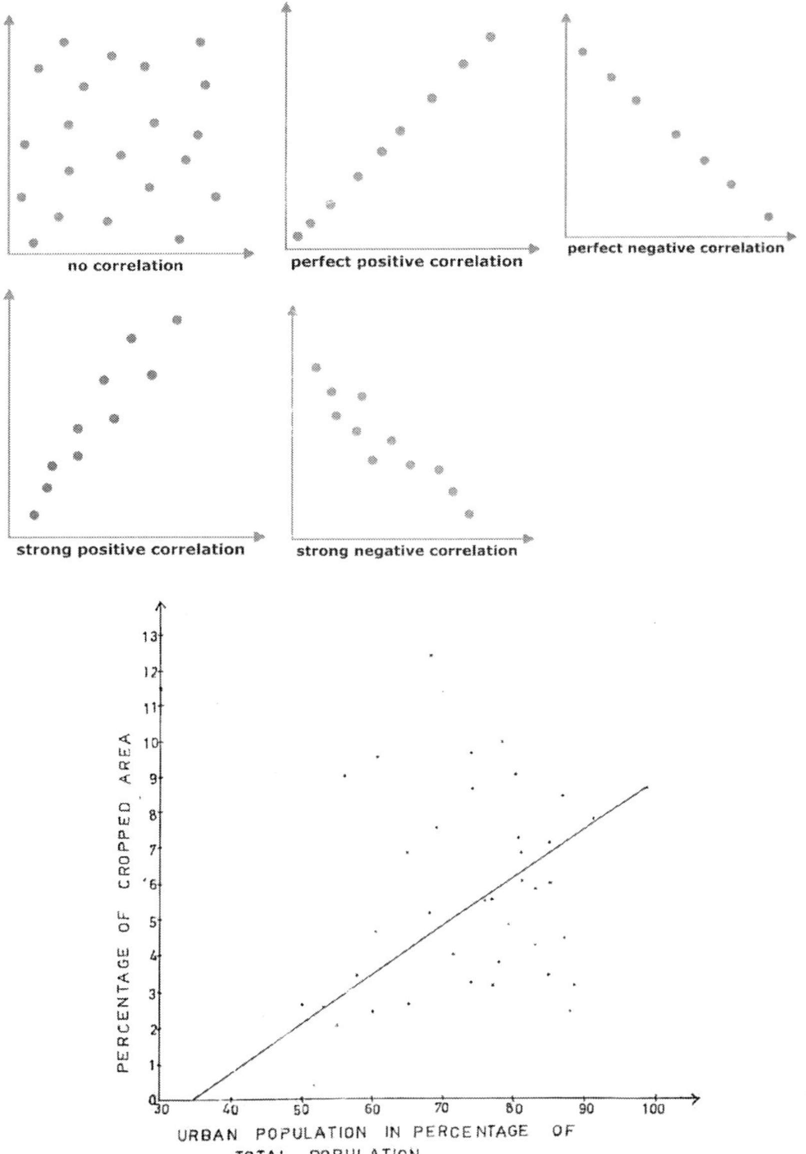

Scatter Diagram Showing Co-relation Between Cropped Area in (%) & Urban Population in (%) of Total Population in Districts of West Bengal

and the other for Y. Each pair of values is plotted as a point on graph paper. The whole set of points higher constitutes a scatter diagram. This method is, of course, not suitable when the number of individuals is very large. In such cases, one may use some three-dimensional analogue of histogram or frequency polygon, constructed on the basis of a bi-variant frequency table.

If the pattern of points or dots on the scatter diagrams shows a linear path diagonally across the graph paper from the bottom left-hand corner to the top right correlation will be *positive*. In other words, the association between the variables is direct indicating thereby that higher values of the other variables and low values are associated with low values.

On the other hand, if the pattern of dots is such as to indicate a straight line from the upper left-hand corner to the bottom right correlation is *negative* i.e., the association is direct, with high values of one variable being associated with low values of the other.

When the dots do not indicate any straight-line tendency but show a concentration around a curved line and correlation is small or will be *zero* or show *no correlation.*

When the dots lie exactly on a straight line, the *correlation is perfect or strong.* The correlation coefficient being +1 or -1 according to the shape of the straight line is positive or negative.

Coefficient of Variation

In order to compare the spread of two or more distributions, the relative measures of variability are used, it is independent of the units of observations and is defined as

$$\text{Coefficient of Variance, CV} = \frac{\sigma}{\bar{x}} \times 100\%$$

Coefficient of Variation is a relative method of measurement of dispersion, absolute measure of dispersion includes range of mean deviation, quartile deviation and standard deviation.

Standard Deviation method is the most important since it takes into account of all the values.

The formula of Coefficient of Variability is CV = S.D. / $\bar{x} \times 100$

Whereas, $\bar{x}$ = mean value

S. D. = Standard Deviation

$= \sqrt{\Sigma(x - \bar{x})^2 / N}$

Data and Calculations For Coefficient of Variability of Rainfall In Burdwan, Suri and Bankura (1950-1979)

	Burdwan			*Suri*			*Bankura*		
Year	*x*	$(x-\bar{x})$	$(x-\bar{x})^2$	*x*	$(x-\bar{x})$	$(x-\bar{x})^2$	*x*	$(x-\bar{x})$	$(x-\bar{x})^2$
1950	48.11	−12.86	165.38	46.43	−6.62	43.82	54.89	−0.99	0.98
1951	46.07	−14.9	222.01	62.05	9	81.00	43.13	−12.75	162.56
1952	57.56	−3.41	11.63	50.63	−2.42	5.86	46.03	−9.85	97.02
1953	44.71	−16.26	264.39	60.52	7.47	55.80	52.4	−3.48	12.11
1954	88.88	27.91	778.97	63.13	10.08	101.61	48.57	−7.31	53.44
1955	78.3	17.33	300.33	57.13	4.08	16.65	61.13	5.25	27.56
1956	44.32	−16.65	277.22	34.84	−18.21	331.60	54.67	−1.21	1.46
1957	65.37	4.4	19.36	60.24	7.19	51.70	50.07	−5.81	33.76
1958	66.15	5.18	26.83	70.31	17.26	297.91	69.02	13.14	172.66
1959	51.14	−9.83	96.63	60.64	7.59	57.61	48.08	−7.8	60.84
1960	47.04	−13.93	194.04	53.17	0.12	0.01	57.5	1.62	2.62
1961	51.33	−9.64	92.93	33.73	−19.32	373.26	33.37	−22.51	506.70
1962	71.99	11.02	121.44	56.6	3.55	12.60	56.6	0.72	0.52

	Burdwan			Suri			Bankura		
Year	*x*	$(x-\bar{x})$	$(x-\bar{x})^2$	*x*	$(x-\bar{x})$	$(x-\bar{x})^2$	*x*	$(x-\bar{x})$	$(x-\bar{x})^2$
1963	55.68	-5.29	27.98	48.48	-4.57	20.88	56.77	0.89	0.79
1964	54.89	-6.08	36.97	46.44	-6.61	43.69	44.71	-11.17	124.77
1965	77.61	16.64	276.89	48.48	-4.57	20.88	56.71	0.83	0.69
1966	66.35	5.38	28.94	46.44	-6.61	43.69	73.49	17.61	310.11
1967	68.25	7.28	53.00	48.03	-5.02	25.20	44.99	-10.89	118.59
1968	56.53	-4.44	19.71	50.6	-2.45	6.00	83.09	27.21	740.38
1969	53.57	-7.4	54.76	91.36	38.31	1467.66	54.67	-1.21	1.46
1970	51.78	-9.19	84.46	50.08	-2.97	8.82	50.03	-5.85	34.22
1971	75.36	14.39	207.07	47.78	-5.27	27.77	64.02	8.14	66.26
1972	68.72	7.75	60.06	46.81	-6.24	38.94	42.08	-13.8	190.44
1973	59.5	-1.47	2.16	48.96	-4.09	16.73	57.08	1.2	1.44
1974	51.31	-9.66	93.32	40.46	-12.59	158.51	62.3	6.42	41.22
1975	70.63	9.66	93.32	57.52	4.47	19.98	49.36	-6.52	42.51
1976	50.68	-10.29	105.88	52.5	-0.55	0.30	80.08	24.2	585.64
1977	63	2.03	4.12	50.09	-2.96	8.76	61.58	5.7	32.49
1978	69.01	8.04	64.64	52.32	-0.73	0.53	46.95	-8.93	79.74
1979	75.36	14.39	207.07	55.8	2.75	7.56	70.68	14.8	219.04

Burdwan	Suri	Bankura
$\Sigma x = 1829.2$ $\Sigma(x-\bar{x})^2 = 3991.52$	$\Sigma x = 1591.57$ $\Sigma(x-\bar{x})^2 = 3345.35$	$\Sigma x = 1674.05$ $\Sigma(x-\bar{x})^2 = 3722.04$
$\bar{x} = 60.973$	$\bar{x} = 53.052$	$\bar{x} = 55.802$
Variance = $\Sigma(x-\bar{x})^2/n = 133.05$	Variance = $\Sigma(x-\bar{x})^2/n = 111.51$	Variance = $\Sigma(x-\bar{x})^2/n = 124.07$
SD = sqrt (Variance) = 11.53	SD = sqrt (Variance) = 10.56	SD = sqrt (Variance) = 11.14
CV = $(SD/\bar{x})*100$ = 18.91%	CV = $(SD/\bar{x})*100$ = 19.90%	CV = $(SD/\bar{x})*100$ = 19.96%

Here, coefficient of variability of rainfall 1950-79, in three districts, is represented by isopleths called equivariables. Isopleths may be used by plotting the places with similar deviations from their average weather conditions. For instance, mean annual figure for any given station does not affect the reliability of rainfall and of those stations. The exact degree of variability is difficult to calculate and various statistical formulae have been derived to represent it. The form of a coefficient of variability is one of such devices. It may be calculated and plotted and isopleths may be drawn. The coefficient of variability is expressed in percentage (%). Higher the coefficient of variability, the nature of distribution is more irregular and uniform distribution is least significant. Coefficient of variability depends on the nature of extreme normal distribution of rainfall and the variability is correlated.

Problem: - To draw the Coefficient of Variability of rainfall of West Bengal from the following given data.

Interpretation

The map of West Bengal shows the Coefficient of Variability of Rainfall from 1950 to 1979 in different districts for about 30 years. This shows that where the rainfall is high, the variability is low, as it can be seen that low variability is found in Medinipur (13.75%), Darjeeling (16.33%), Hooghly (16.77%) and Burdwan (18.91%). High variability is found near Balurghat (33.50%), Cooch Behar (26.23%), Howrah (21.13%). In Cooch Behar (26.23%), there is little rainfall but in Balurghat and Howrah due to absence of vegetation and city/urban influence with smoke and dust, rain bearing clouds have drifted apart. Due to the formation of heat islands in those urbanized areas, especially in Howrah and adjoining areas, the formation of clouds is inhibited. But in areas like

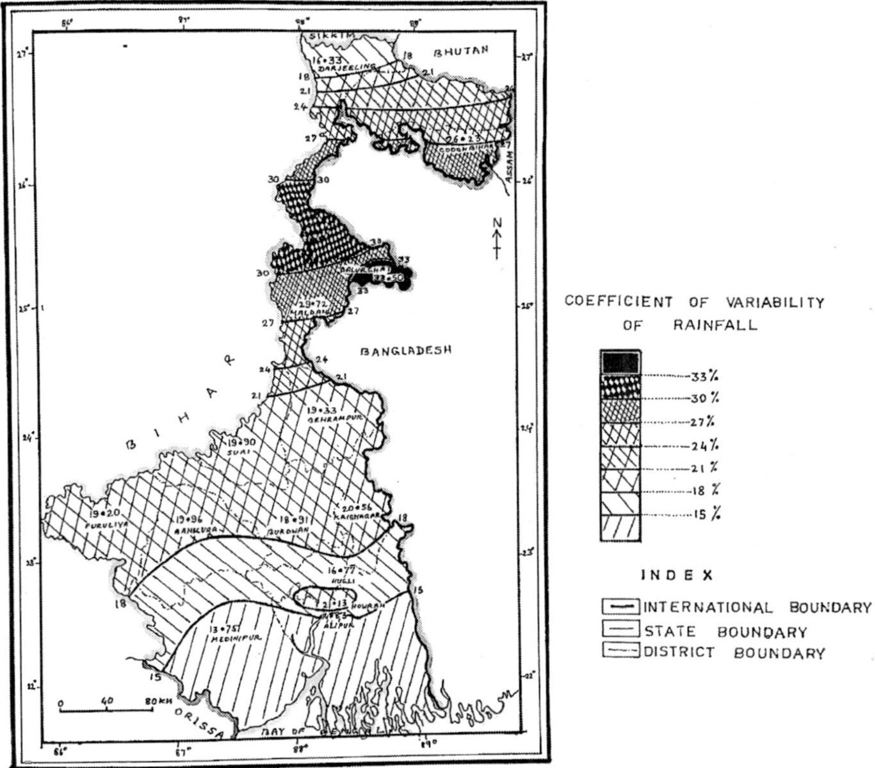

Coefficient of Variability of Rainfall in West Bengal 1950–79

Suri (19.90%), Bankura (19.96%), Berhampur (19.33%) and Purulia (19.20%) i.e., the Western plateau and Rarh region of West Bengal the coefficient of variability of rainfall though moderate but has an increasing trend. The amount of rainfall is decreasing, primarily, due to its geographical conditions accentuated by the high rate of urbanisation and industrialisation which simply impede rain formation. The existence of natural vegetation which is the prime factor of causing rain are being continuously cleared for human settlements and so there is marked variation in the decrease of rainfall.

PRACTICUM

15

Unit Planning for School Geography

Introduction

Unit planning serves as a roadmap for educators, guiding them in navigating the diverse terrain of geographical concepts, skills, and inquiries. By meticulously crafting each unit through the processes of Preparation, Presentation, Association, Generalisation, and Practical Application, educators create cohesive learning experiences that engage learners and foster deep understanding and critical thinking skills. In general, unit planning in geography education has evolved and kept to the updates and demands of society.

Understanding the Curriculum

In secondary and senior secondary geography education, curriculum standards emphasize geographic literacy, spatial reasoning, and critical thinking skills. For example, learning objectives may include understanding processes shaping Earth's landforms and analyzing patterns of population distribution and migration. Integration with other school disciplines enhances learners' holistic understanding of geographic concepts.

Analyzing Learner Needs

Before designing units, educators assess learners' prior knowledge, learning styles, and abilities. For example, a pre-test on weather and climate comprehension gauges learners' understanding, while accommodating diverse learning styles through visual aids and hands-on activities ensures inclusivity. Strategies for varying proficiency levels, such as differentiated instruction, promote equitable learning experiences.

Setting Clear Learning Goals

Clear and measurable learning objectives ensure alignment with curriculum standards and guide instruction effectively. For instance, in a unit on migration, learners analyze push and pull factors influencing migration patterns, aligning with specific standards related to population dynamics and human geography. Content-based and skill-based goals promote a balanced approach to instruction, fostering comprehensive understanding and critical thinking skills.

Selecting Relevant Content

Relevant content selection involves identifying core concepts, themes, and topics that resonate with learners' experiences and interests. For example, exploring urbanization encompasses core concepts such as population growth and land-use patterns, integrating contemporary issues like climate change's impact on urban environments. Interdisciplinary connections with subjects like history and economics enrich learners' understanding of geographical phenomena.

Designing Engaging Learning Activities

Engaging learning activities cater to diverse learning needs through hands-on experiences, fieldwork, technology integration, and group work. For instance, hands-on activities like creating topographic maps immerse learners in practical applications of geographical concepts. Fieldwork, such as conducting surveys or exploring geographical features, connects classroom learning with real-world experiences.

Assessing Learner Learning

Assessment strategies include formative and summative methods that cater to diverse strengths and preferences. Formative assessments, like quizzes and concept maps, provide ongoing feedback to guide instruction. Summative assessments, such as projects and presentations, evaluate mastery of content and skills. Authentic assessment tasks, mirroring real-world applications, enhance relevance and engagement. Diagnostic tests followed by remedial techniques enable a learner to perform and thereby appear for placement tests for higher knowledge acquisition.

Differentiating Instruction

Differentiated instruction addresses diverse learning needs through scaffolding, flexible grouping, differentiated assignments, and assistive technologies. For example, scaffolding techniques gradually remove support as learners master concepts, ensuring inclusivity and promoting independence. Flexible grouping allows educators to tailor instruction to learners' abilities and interests, fostering collaboration and peer learning.

Incorporating Reflection and Feedback

Reflection and feedback promote metacognitive skills and guide learner progress towards deeper understanding. Opportunities for reflection, like journaling and think-pair-share discussions, encourage self-awareness and ownership of learning. Timely and constructive feedback, aligned with learning objectives, guides improvement and fosters continuous learning. Encouraging self-assessment and goal setting empowers learners to track progress and take ownership of their learning journey.

In the context of unit planning, discussing both merits and demerits provides a comprehensive understanding. It recognizes the strengths of unit planning and acknowledges potential drawbacks ensuring a balanced perspective for educators.

Merits of Unit Planning

Alignment: Unit planning ensures all geography activities match specific learning goals, like analyzing weather patterns or understanding cultural landscapes.

Coherence: Structured unit plans in geography allow learners to understand and progress logically and ensure better retention say, studying landforms to exploring the impact of human activity on ecosystems.

Demerits of Unit Planning

Time-consuming: Crafting detailed geography unit plans requires significant time investment, diverting educators from other vital tasks like creating interactive maps or conducting fieldwork.

Inflexibility: Rigid adherence to geography unit plans might hinder educators from seizing impromptu teaching moments, like discussing current events shaping global migration patterns.

Conclusion

Effective unit planning and instructional strategies are essential for enhancing geography education. It creates a meaningful learning experience that empowers learners to become informed global citizens. As geography education continues to evolve, educators must embrace continuous reflection and refinement of unit planning practices to meet the evolving needs of learners in society and the discipline.

References: List of recommended further readings, resources, and websites for further exploration of specific required content can be done for enhancing geography education.

FORMAT FOR UNIT PLANNING -- Lesson No.

Class:
Unit/Sub-unit:
Total No. of Period(s) needed for Unit/Sub-unit(s):
Duration of each Period:
Time Duration needed for Evaluation:
Time Duration needed for Remedial Teaching:

Sl. No.	*Topic- Unit / subunit & No. of Period(s)*	*Specifying Sub-concept(s) with specific learning objectives*	*Appropriate Methods of Teaching (as required for General and Special needs)*	*Teaching-Learning Material(s)/ resources*	*Strategies of Teaching-Learning Activities - Teacher/ Learners*	*Evaluation/ Assessment Strategies*

RATING SCALES FOR EVALUATING INTERNSHIP ACTIVITIES
Rating scale for Evaluation of Activity: Preparation of Unit Plan

Teaching Subject: **Class/Section:**
Name of the Candidate: **Roll/ Regn No./Enrolment No.:**
Name of the school:

Rating is to be done according to the following criteria:

Criteria	**Rating**				
Unit plan has covered the required number of sub-units	1	2	3	4	5
Stated the learning objectives of the unit	1	2	3	4	5
Sequenced the contents as per the learning objectives of the unit	1	2	3	4	5
Suitable methods and media have been suggested	1	2	3	4	5
Suitable evaluation methods have been suggested	1	2	3	4	5
Prepared appropriate unit-end questions for evaluation	1	2	3	4	5
Reflections included in the Report	1	2	3	4	5
Quality of Presentation in the Report	1	2	3	4	5

Note: Rating Score: 1: Very Poor, 2: Poor, 3: Average, 4: Good, 5: Very Good

- Any specific material used or adapted for the learners with special needs..............
- Remarks /suggestations ....

Overall rating: **Signature of the Evaluator with Date**

16

Lesson Planning /Learning Design: Geography

Introduction

It is primarily important to mention that the progressive term of 'Lesson Plan' is taken as 'Learning Design'. Effective classroom teaching highly depends on the proper design of a lesson. Prior knowledge of the subject matter and mode of delivery helps a teacher to do justice to the students. Experienced teachers may not write learning designs in detail as they are quite efficient with the steps, but beginner teachers are always advised to be careful in preparing a detailed learning design for their success as a teacher in the class.

A learning design is a plan of action. It includes the working philosophy of the teacher, his knowledge of philosophy, his information about and understanding of his pupils, his comprehension of the objectives of education, his knowledge of the materials to be taught, and his ability to utilize effective methods or procedures. In the words of Bossing (cf. Sharma and Chandra, 2003: p-506), 'Lesson plan is the title given to the statement of the achievements to be realized and the specific means by which these are to be attained as a result of the activities engaged in, during the period'.

Learning design is a brief outline of the main points of the lesson to be covered by the teacher in a specified period for the realization of specified objectives. It indicates clearly what already has been done, what the pupils are to do, how the pupils are to be engaged in various activities, and what activities are to be pursued. It is a clear and precise statement of the aims and purposes of the lesson and the various techniques and devices used by the teacher. Daily learning design involves defining the objectives, selecting and arranging subject matter, and determining the methods or procedure.

In preparing a learning design a teacher has to consider the following points: i. Broader objectives of the subject ii. Instructional objectives of lesson iii. Organization of subject matter to be covered in the lesson for achieving the stipulated objectives iv. The decision about the way of presenting the subject matter, teaching strategies and tactics, classroom interaction, and management v. Appropriate provision for evaluation and feedback.

Importance of Learning Design

A good learning design can perform a number of functions as follows:

1. Learning design helps a teacher to define his aims and objectives more clearly. The teacher thinks of ways and means with which he can realize his aims most effectively.
2. It helps a teacher in the selection and organization of subject matter, materials, and activities, according to the abilities, aptitudes, and developmental levels of students.

3. Learning design helps a teacher in the selection of the most effective teaching procedure, which will lead to the modification of pupils' attitudes, habits, and information in a desirable direction.
4. Learning design helps a teacher to evaluate the outcomes of instructions. Evaluation is a very important part of the teaching-learning process. Evaluation is possible if definite aims and objectives are kept in mind.
5. Learning design helps a teacher to be systematic and prevents wastage. It saves him from haphazard teaching and repetitions are avoided.
6. Learning design establishes the proper connection between the different lessons and the units/subunits of study. Thus, it encourages continuity in the teaching process.
7. Learning design provides adequate lesson summaries and ensures a definite assignment for the class which helps in developing insight and understanding.
8. Learning design imparts confidence and self-reliance to the teacher. It enables the teacher to enter the classroom without anxiety and foresees the difficulties that are likely to arise and get prepared to deal.
9. It facilitates the teacher a series of plans well in advance to progressively modify children's attitudes, habits, and abilities towards desirable directions.

Principles of Learning Design in Geography

1. Principle of Writing the Learning design: A good learning design should be preferably written and should not remain at the oral or mental stage for the management of both theoretical and practical/project sections of Geography.
2. Principle of Statement of Objectives: A Geography learning design must contain both general and specific objectives which should be clearly stated.
3. Principle of Selection of Suitable Subject Matter: The Geography content of the lesson plan should be selected according to the needs, interests, abilities, and levels of the pupils (as specified in the syllabus).
4. Principle of Orderly Presentation: The subject matter should be presented in an orderly, organized, and effective manner with the appropriate geographical instructional aids and techniques. The effective presentation includes-
   a) Principles of Co-Relation: The subject matter which is following, need to be correlated with previous knowledge by the teacher. Moreover, it should be co-related with various other disciplines for best integration. It will make the lesson interesting and meaningful.
   b) Principle of Teaching Techniques: The plan should indicate the teaching techniques to be used by the teacher-how lesson is to be presented, what method is to be followed, what questions are to be asked, and what illustrations are to be used. It should clearly indicate the teaching-learning aids to be used by the teacher.
   c) Principle of Use of Chalk-Board: Learning design should clearly indicate chalkboard work done by the teacher. It should point out the diagrams and sketches which are to be drawn on the chalkboard.
   d) Principle of Child's Activity and Participation: Learning design should point out the child's activities and participation which are to be made in the classroom. Pupils should be encouraged to ask questions, solve problems and perform demonstrations/experiments.

5. Principle of Flexibility: Learning design should be planned so that lesson transaction is successful for all learners in an inclusive set-up.
6. Principle of Evaluation: Learning design should indicate evaluation activities. The evaluation exercise must consist of short answers and objective-type questions.
7. Principle of Assignment: Geography Learning design should include assignments based on contemporary issues for pupils.
8. Principle of Bibliography: Learning design should indicate selected references-printed/online.
9. Principle of Duration: The learning design should indicate the time within which the lesson is to be delivered. It must be balanced and well pervasive.
10. Principle of Age: The learning design should point out the target group i.e., age/grade of pupils. The lesson is to be planned for a particular stage of development and the mental level of the pupils.
11. Principle of Making Summary: The learning design should have an outline or summary of the entire lesson which may be also developed on the chalkboard with the help of the students.

Qualities of Geography Teacher for Preparation of Learning Design

The requirements of a geography teacher for the creation of effective learning designs can be outlined as follows:

- A teacher should hold an in-depth knowledge of the environmental conditions and also possess the managerial capacity of the students (individually and collectively) in an inclusive classroom
- Ability to make the students a thorough comprehension or consolidation to learn the lessons based on both the physical and psychological abilities of the students
- A mastery of the subject, motivating skill for the integration of co-scholastic activities along with the lesson transaction
- Efficiency in the appropriate handling of different instructional aids and techniques of teaching for flexible class management
- Ability to meet the specific needs of the diversified learners in the class in accordance with the topic

Steps to Prepare a Learning Design of Geography

There are different approaches to writing a learning design. H. C. Morrison and B. S. Bloom had given their own schemes on the basis of the unit approach and evaluation approach respectively. In India, the Regional College of Education, Mysore (RCEM) had also developed another model of the lesson plan. Herbart's model is, however, still the most popular one, and it is inclusive of the following five steps:

Preparation> Presentation> Comparison and Association> Generalization> Application Herbart's model was briefly described under the following heads: 1) *Information of identification* 2) *Instructional objectives 3) Previous knowledge 4) Announcement of the aim 5) Presentation with Teaching aids 6) Practice work/Homework 7) Evaluating a lesson*

A format to guide the students in constructing the learning design has been provided below:

Basic Format for Learning Design

Name of the School: Class: Duration: Date: Teacher's Name:	Subject: Lesson Unit: Sub-units: Today's Lesson:

Learning Goals/Objectives

[Goal- Statements based on Revised Bloom's Taxonomy]

Learner will be able to.... *Factual Knowledge* *Procedural Knowledge*	*Conceptual Knowledge* *Metacognition*
Remembering	←
Understanding/ Comprehending	←
Applying	←
Analyzing	←
Evaluating	←
Creating	←

Analyze Learners and Context

[General characteristics of the learners in terms of entry-level behaviour, to be ascertained by putting relevant questions]

To know the general characteristics and entry-level behavior of the learners, the following questions may be asked with respect to the present context.

Develop & Select Learning Materials

[Text Books and references; Teaching Learning Materials]

Learning Strategies

Learning Areas	Related Strategy

Design for Evaluation

[Test items/ work sheets/models/ experiments/ creative writings/ creative performances etc.]

Diagnosing the weaknesses for remedial class (if any)

[Teacher will diagnose the weaknesses after the evaluation]

N.B. Factual Knowledge may be described as the basic information about a particular subject or discipline that students must be acquainted with. *Conceptual Knowledge-* refers to the knowledge of, or understanding of concepts, principles, theories, models, classifications, etc. *Procedural Knowledge-*also known as imperative knowledge, is the knowledge exercised in the performance of some task. *Metacognition-* "thinking about thinking", "knowing about knowing", becoming "aware of one's awareness" and higher-order thinking skills.

Remedial Teaching

This is the most important and final phase of a lecture because the transaction of ideas and information between the teacher and the learner takes place to improve the learning skill and rectify a particular problem. The teacher explains and recollects the concepts and principles, and provides facts, data, figures, etc., to the learners very briefly. In order to explain the context matter, the teacher once again cites various examples, uses various communication aids, uses analogies and illustrations, or by writing them again on the blackboard, etc. The teacher, when required, adopts different non-verbal communication techniques such as gestures, postures, etc. to facilitate the teaching activity. During this phase, the teacher should be cautious of the lecture, otherwise, it becomes ineffective and a repeated one. In this concluding phase of the lecture, the teacher usually recollects and summarizes whatever has been covered during the course of the lecture and so one might also ask a few questions on the content matter, in order to evaluate the understanding of the lecture by the students. Through these questions, the teacher gets to know the learning difficulties of students and accordingly modifies the teaching. This reflective teaching identifies learners with diverse learning needs. The teacher also gives some home tasks to the students.

Application and Limitations of Learning Design in Classroom Situation

Effective classroom teaching highly depends on the proper planning of a lesson. Prior knowledge of the class, subject matter, and mode of delivery helps a teacher to do justice to the diversified learners. Experienced teachers may not write learning designs in detail as they are quite efficient with the steps involving a learning design, but beginner teachers are always advised to be careful in preparing a detailed learning design for their success as a teacher. However, learning designs also have certain limitations, as outlined below:

1. A learning design is actually a plan of action. But sometimes, it requires goals and objectives of teaching through proper selection of subject matter, teaching procedure, methods of evaluation, and other planning activities in accordance with the situation. This requires precision and efficiency on the part of the teacher in order to create a scope for self-evaluation.
2. A learning design is a well-defined track ensuring steady progress and a definite outcome of the teaching-learning process that needs to be abided by the subject-teacher. It should be formulated in a manner so that interest is generated and they are keen to assimilate the subject. However, often the real-world situation may not be conducive to executing a well-prepared learning design.
3. Learning design should be done in such a way that it facilitates the teacher to progressively modify children's attitudes, habits, and abilities towards desirable directions. Vital points, quotations, and 'telling phrases' on current geographic topics need to be readily used to make the class abreast of the present world situation.
4. Learning designs need to be well-knitted and interconnected so as to avoid repetition and monotony. Learning design needs an appreciable set, to meet the general and specific needs of the learners in an inclusive class.

5. Time management and the need for an inclusive situation should adhere to the flexibility that exists in the curriculum transaction. There should be no rigidity in handling different aids and techniques of teaching particularly in disseminating the subject matter in the class.
6. It is of absolute necessity that the teacher should possess an intimate knowledge of environmental conditions and have an academic influence on the students present in the class.
7. Questions framed should be relevant and applicable for all learners possessing individual differences.
8. Care to incorporate a list of further reading should be included as the essential part of the learning design.

Learning Design

Name of the School: Class: VIII **Time:** 40 minutes **Date:** **Name of the Teacher:**	**Subject:** Geography **Unit:** Oceania **Sub-units:** • Oceania expedition • Physical and political divisions * Murray-Darling basin * **Today's Lesson:** Murray-Darling basin

A. Learning Goals/Objectives (Based on Revised Bloom's Taxonomy)

Domain/Area	Goals/Objectives
1. Remembering	The learners will be able to identify and delineate the location of Murray-Darling basin. (FK) The learners will be able to state the extent and volume of Murray-Darling river. (FK)
2. Understanding	The learners will be able to interpret the significance of Murray-Darling river. (CK) The learners will be able to explain the flow and direction of Murray-Darling river. (PK)
3. Applying	The learners will be able to describe the agricultural practices and cattle-rearing activities and their developmental trends through time. (FK) The learners will be able to interpret the cause of the growth of temperate grasslands (Downs). (CK)
4. Analyzing	The learners will be able to analyse the economic activities prevalent in the area. (CK) The learners will be able to reason out the favourable causes of the various crops that are grown in various parts of the Murray-Darling basin. (CA)
5. Evaluating	The learners will be able to judge the contribution of Murray-Darling river which has facilitated the region to develop economically. (FK) The learners will be able to evaluate how Murray-Darling has affected in the growth of the population in the region. (FK)
6. Creating	The learners will be able to draw the flow of Murray-Darling in Oceania. (MK)

B. Analyzing Learners and Contexts

To ascertain the general characteristics of learners in terms of entry-level, the following questions are to be put up—

- Why is the term 'Oceania' used?
- What are the major rivers of Oceania?
- Which area/physical division of Australia comprises the Murray-Darling valley?
- Why is Murray-Darling a significant river system?
- Why are Downs famous?
- What economic activities are possible based on Oceania's physical characteristics?

C. Developing and Selecting Learning Materials

General Learning Aids:

1. A prescribed Geography textbook of the concerned Class VIII
2. Chalk-duster and blackboard (part of the classroom).

Specific Learning Aids: Achart prepared showing the location of Murray-Darling basin, and a flashcard showing the directional flow of Murray-Darling.

D. Learning Strategies

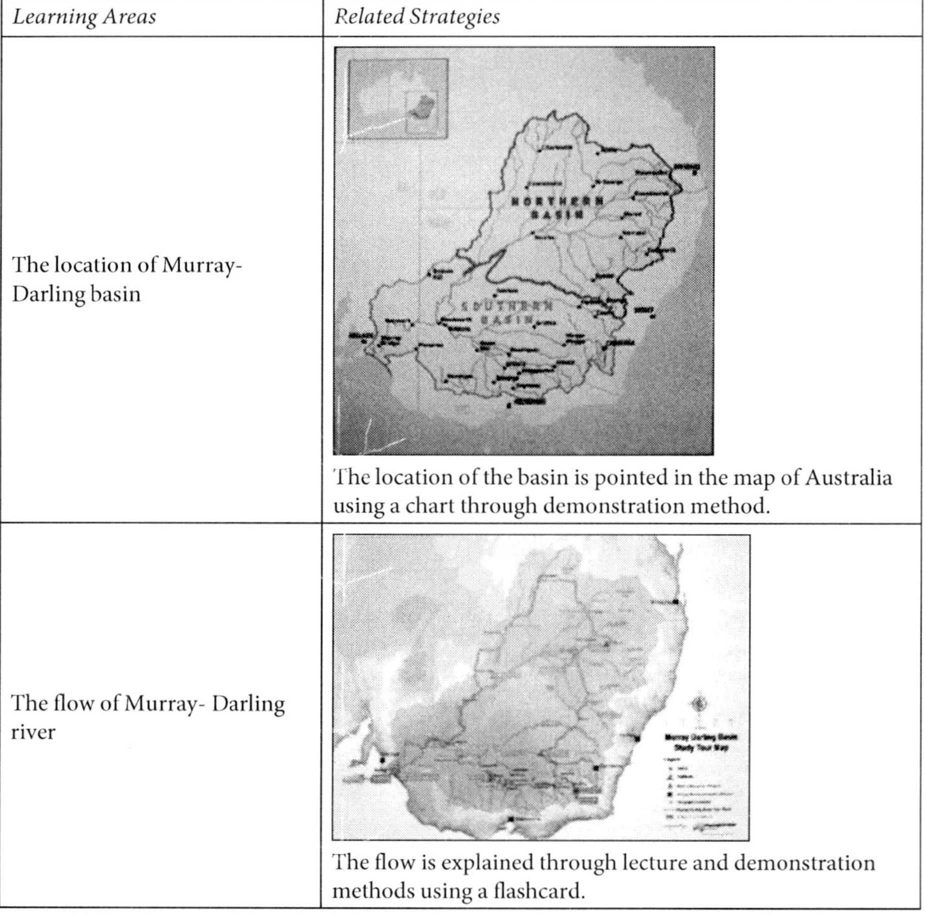

Learning Areas	*Related Strategies*
The location of Murray-Darling basin	The location of the basin is pointed in the map of Australia using a chart through demonstration method.
The flow of Murray- Darling river	The flow is explained through lecture and demonstration methods using a flashcard.

Economic activities and areas under the economic activities	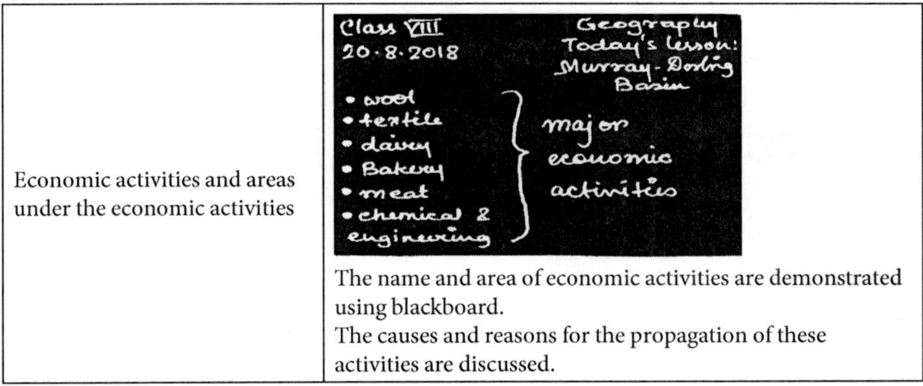 The name and area of economic activities are demonstrated using blackboard. The causes and reasons for the propagation of these activities are discussed.

E. Designing for Evaluation

In order to formulate the evaluation procedure, the teacher would design the formative and summative evaluation measures.

Formative Evaluation: To check the comprehensibility of the lesson concept, the teacher monitors the learning effectiveness by drawing attention and concentration, awakening motivation through goal-oriented behaviours, and promoting learner interaction through collaborative and socialization techniques.

Summative Evaluation: After the completion of the lesson, the teacher would conduct the performing tasks through the preparation of worksheets or experiments.

WORKSHEET (given in the class)

1. **Fill in the blanks:**
   (a) The grassland in Murray-Darling basin is _______.
   (b) _______ is called 'Silver City'.
   (c) The most important animal producing wool and fur is _______.
2. **State 'True' or 'False':**
   (a) The slope of Murray-Darling basin is towards west.
   (b) The Murray-Darling rivers have their confluence in Wentworth.
3. What type of climate prevails in Murray-Darling basin?
4. State the most populous area of Australia.

As a part of Homework:

A Project-based Learning activity on the '*Significance of the Murray-Darling basin in the lives of the people*' would be given by the geography teacher to the learners.

F. Diagnosing the Weakness for Remedial Class (if any)

In order to support/assist the learners with their learning difficulties, the teacher would formulate/ organise, the remedial class for them.

The aspect of climate on the basin environment was confused.	The role of climate on the basin and its relationship to be clarified through lecture and discussion methods and the location map on chart can also be used.

17

Pedagogical Analysis of Content: Geography

Definition and Purpose

Pedagogical analysis of content refers to breaking up the content of a particular topic of discussion into smaller parts, e.g. into units or sub-units for effective or scientific teaching. The teaching can therefore become meaningful, serving the intended purpose. A difference between content analysis and pedagogical analysis can be made, and this is important for a better understanding of the two. However, it is to be noted that the difference is solely dependent on the purpose or objective. In *content analysis* the objective is to break the whole subject matter into smaller parts, properly organized in sequential and meaningful order; whereas in *pedagogical analysis* similar partition of the subject matter is done following a suitable methodology for the purpose of scientific teaching. Effective teaching is defined as 'the ways and means provided to or utilized by a teacher for managing the task of teaching as smoothly and effectively as possible by involving his least efforts for the best possible outcomes' (Mangal and Mangal, 2008). Any course or syllabus comprises a number of chapters or units. Each unit consists of a quantity of knowledge that represents a particular theme. This body of knowledge has to be transacted in the class so that the objectives of teaching the unit are achieved successfully by the students. This necessitates that the unit/content is planned in a systematic manner to achieve the objectives of the teaching-learning process. The main objective of a unit plan is to manage student learning in that particular content. It represents the sequencing of learning activities of the content and the supporting learning materials that are required for the transaction of learning experiences. So, the content analysis serves the following purposes: 1) Smooth and orderly teaching 2) Maximum output with the least effort 3) Conceptual clarity among both the teachers and students 4) Effective management of time and skill and other necessary resources 5) Effective prior programming of teaching methods and selection of suitable teaching aids.

Operational Steps of Pedagogical Analysis and the Role of A Geography Teacher

In today's modern forms of teaching geography, both pre-service and in-service learners should be in a position to evidently: a) Acquire knowledge by themselves, understand issues, and be able to apply the knowledge and skills in life b) Build on their own efforts, gain knowledge, understanding, and experiences by working on activities and assignments supervised by the lecturers/teachers and engaging often in fieldwork activities and discovery expeditions in order to perfect their working skills for practical learning. c) The process of learning must model crucial aspects of critical thinking, reasoning, observations, and research. The learners should have ideas of what they want to learn and ought to establish facts and understanding of the topics coupled with a philosophical understanding of how the content will assist one in developing a deep understanding of issues learned is important. This will result in the learner developing action plans for handling issues in life.

The pedagogy involves interrelated discourse of four major elements of the teaching-learning process, namely- (1) Instructional objectives; (2) Content material and learning experiences; (3) Tools/ aids and techniques of teaching; and (4) Evaluation methods.

Content analysis is based on the interdependence of these four aspects of the teaching-learning process, and therefore it includes the following steps:

(a) ***Analysis of the content of the topic to be taught***
This requires a thorough and complete understanding of the content-material prescribed in the Geography syllabus of both secondary/senior secondary. So, a good textbook may be consulted to take note of the important points of the content.

(b) ***Setting the instructional objectives in behavioural terms***
After going through the teaching-learning process on a selected geographical topic, the students are required to show some definite types of behavioural changes in them. This can be identified with the help of certain questions/ tests based on various domains of educational objectives.

(c) ***Enlistment of the teaching aids and methods relevant to the contents of the topic***
The list of geographical teaching materials, methods, teaching aids, and activities should conform to the relevant conditions. Besides the textbook study, there can be demonstrations, exhibitions, or projections of the material; a teaching method may be selected from different alternatives for the benefit of inclusive learners.

(d) ***Selection of the suitable technique for proper evaluation***
The evaluation of the geographical learning outcome may be based on CCE/ competency-based or on written/ oral tests relevant to the theoretical/practical concept-formation. Other evaluation techniques like observation, rating scale, and interview may also be adopted.

Application and Limitations of Pedagogical Analysis in Classroom Situation

Pedagogical analysis of content refers to breaking up the content into units or sub-units for scientific and retentive teaching. This integrated teaching can become meaningful for serving the intended purpose. Although it has many rewarding applications, it has certain limitations too, as mentioned below:

1. Experience shows that class teaching would not be smooth and orderly with the help of pedagogical analysis only, because it obviously depends upon the particular classroom situation. In accordance with the demanding situation, the orientation of classroom teaching may be regulated. Therefore, strict adherence to the pre-determined pedagogical analysis of content is not feasible.
2. It is not always necessary that there would be maximum output with the pedagogical analysis because the handling of the classroom teaching requires appropriate mastery of the subject and conceptual clarity by the teacher concerned.
3. Effective prior programming of teaching methods and selection of suitable learning-aids demand efficiency on the part of the teacher, because it particularly depends upon the class size and the socio-economic background of the learners; and also, upon the infrastructural facilities available in the school.
4. Elaborate plans on teaching strategy, preparation of test items, table of specification, and worksheet does not necessarily lead to the best performances

in the class. The teacher may have the ability to prepare such items but may lack in-depth knowledge of the matter and its purpose. The trainee-teacher may work in a very mechanical manner and thereby the entire system may become futile.

5. Emphasis on the appropriate break-up and detailing of the sub-units obviously depends upon the class, the psychology of the learners, and the nature of the selected topic. This requires precision, understandability, and presentability on the part of the teacher.
6. Pedagogical analyses on a particular topic/unit/sub-unit and class are prepared beforehand on a very standard set of norms. But on the standard, the psychological capability of the students and the availability of the infrastructural facilities may vary from school to school. This may lead to the non-applicability of the entire procedure and everything might go in vain.
7. Too much time is spent preparing a well-planned pedagogy. And that's why the training procedure for the completion of the required number of pedagogies on a specified subject in such a short span of time is, to a great extent, limited.
8. Pedagogical analysis lays down a general concept of a teaching-learning programme for every student. But attention should be duly paid to serving all categories of students who may be gifted, average, and slow learners.
9. Unnecessary elaboration and exaggeration of the sub-units in the pedagogical analysis especially by inexperienced teachers may distort the spontaneous learning process in the classroom.

PEDAGOGICAL ANALYSIS

Breaking of Units into Sub-Units with no. of Periods

Class: X
Subject: Geography
Unit: Atmosphere

Sub-units:	*Number of Periods*
• Concept and Definition of Atmosphere	
• Composition of Atmosphere	1
• Structure of Atmosphere	3
• Heat, Temperature, and Global Warming	6
• Pressure Belts and Winds	7
• Humidity and Precipitation	4
• Major Climatic Regions of the World	3
• Remedial Class	1
Total Periods	25

Sub-unit Selected: Structure of Atmosphere

Previous Knowledge

1. The learners are aware of different elements of weather and climate.
2. Pupils have an idea of the concept of the atmosphere.
3. Learners know the varied components of the atmosphere.
4. Learners have cognizance about various proportions of gasses that exist in the atmosphere.

INSTRUCTIONAL OBJECTIVES IN BEHAVIOURAL TERMS

Domain/ Area	*Goals/Objectives*
1. **Remembering**	a. **Name** the hottest layer of the atmosphere. (FK) b. **State** the significance of the presence of the Ozone layer in the atmosphere. (CK) c. **Underline** the incorrect word of the statement given below: *The increase in temperature with an increase in height is known as the Normal Lapse Rate.* (PK) d. **Make facts- chart** on Thermosphere. (MK)
2. **Understanding**	a. **Interpret** the characteristic features of the Exosphere. (FK) b. **Distinguish** between the Homosphere and Heterosphere. (CK) c. **Discuss** the procedure of formation of auroras in the Ionosphere. (PK) d. **Find more information** on the causes behind Ozone Depletion. (MK)
3. **Applying**	a. **Investigate** the CFC generation sources from the household sector. (FK) b. **Illustrate** with an example the detrimental effects of depletion of the ozone layer. (CK) c. **Put together your knowledge** of the Thermosphere being responsible for enabling radio broadcasting and wireless communication. (PK) d. **Change** the following into the correct order of sequence from the surface of the earth: *Exosphere, Troposphere, Thermosphere, Stratosphere, and Mesosphere.* (MK)
4. **Analyzing**	a. **Explain** the reason for meteors' movements slowing down and burning up in the Mesosphere. (FK) b. **Compare and contrast** between the Troposphere and Stratosphere. (CK) c. **Analyze** the reason behind the thermosphere being the hottest layer in the atmosphere. (PK) d. **Investigate** the reason behind atmospheric pressure being lowest in the Exosphere. (MK)
5. **Evaluating**	a. **Discuss** the role of the Troposphere as a weather-making layer in the atmosphere. (FK) b. **Find the importance** of the presence of the ozone layer in the atmosphere. (CK) c. **Assess** the reason for the stratosphere providing ideal conditions for flying. (PK) d. **Validate** the need or importance of the presence of the Thermosphere and its positive impacts on human lives. (MK)
6. **Creating**	a. **Formulate** a chart showing different layers of the atmosphere with proper labeling. (FK) b. **Combine** the significance of each layer stating their importance in the earth's atmosphere in a write-up. (CK) c. **Create** a diagram of the depletion of the ozone layer and label it. (PK) d. **Prepare** a block diagram on the division of the atmospheric layer based on chemical composition and variation in temperature. (MK)

[Note: - FK- Factual Knowledge; CK- Conceptual Knowledge; PK- Procedural Knowledge; MK- Meta-cognitive Knowledge]

SUB-UNIT WISE CONCEPTS

Sub-Unit	*Broad Concepts*	*Brief Summary*
Structure of Atmosphere	**Concept 1:** Layers of Atmosphere according to the chemical composition	• The atmosphere is made up of a mixture of gasses that form a layer surrounding the earth. Water vapour and dust particles are other constituents of the atmosphere. • According to its chemical composition, the atmosphere can be divided into two layers, *i.e.* (a) The proportion of gasses varies greatly with the layers of height. Up to the height of 100 km, the composition of the atmospheric layer remains almost the same and it is popularly known as ***Homosphere.*** (b) Above this layer, the percentage of gasses varies and shows heterogeneous character; this upper layer nearly up to 10,000 km in height is known as the ***Heterosphere.***
	Concept 2: Layers of Atmosphere according to variation in temperature	• Based on temperature differences atmosphere can be divided into five concentric layers, *i.e.* • **The troposphere** is the lower-most layer of the atmosphere, nearest to the Earth. All daily weather changes take place in this layer (extends up to 14 km from the surface around the Equator and about 8 km around the poles.) • **The stratosphere** lies above the Troposphere (extends up to a height of 60 km) It provides ideal conditions for flying Jet planes and Commercial aircraft. • At the **Mesosphere,** the meteors that come towards the earth get burnt in this layer (which extends up to a height of 80 km above the earth's surface.) • In **Thermosphere,** ionization occurs in this layer, hence, this layer is also known as the Ionosphere (it extends from about 90 km to about 500-1000 km above our planet.) • **The exosphere**, the uppermost layer of the atmosphere (exceeding 700 km in height) merges into space gradually. • In the case of the Troposphere and Mesosphere, the Normal Lapse Rate of temperature operates where temperature decreases with increasing height but in the case of the other three layers i.e., Stratosphere, Thermosphere, and Exosphere, inversion of temperature is observed in which, with increasing height the temperature increases.

TEACHING - LEARNING STRATEGIES

The students will be able to understand the concepts of the chosen subunit by applying the following strategies:

Learning Areas	*Related Strategies*
Concept 1: Layers of the Atmosphere according to chemical composition.	• Through the Lecture and Demonstration methods, the layers of the atmosphere according to chemical composition will be introduced. • The trainee-teacher will draw a tree-chart to explain the concept of the Homosphere and Heterosphere and she/he will write down important features of these layers on the blackboard.

Learning Areas	*Related Strategies*

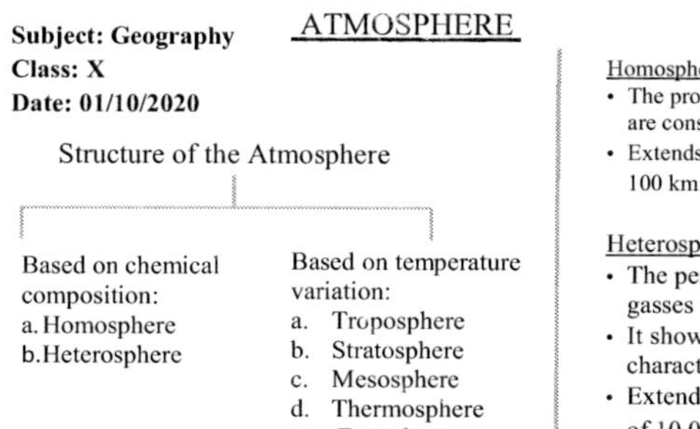

Black Board Work No. 1: Structure of Atmosphere

- The structure of the atmosphere based on chemical composition is taught using a Chart, namely *"Layers of Atmosphere Based on Chemical Composition"*

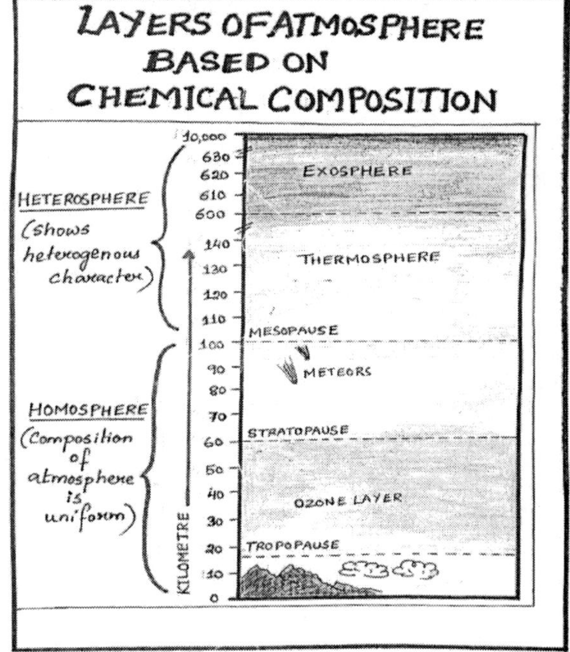

Figure No. 1: Chart Showing Layers of Atmosphere

Concept 2: Layers of Atmosphere according to temperature variation

- Demonstration, interactive, and lecture methods will be used to teach pupils about different layers of the atmosphere according to temperature variation.
- The trainee-teacher will teach the characteristics of different layers of the atmosphere, their altitudinal ranges, and temperature variation through the lecture method and relevant features of each layer will be written down on the blackboard.

Learning Areas	*Related Strategies*

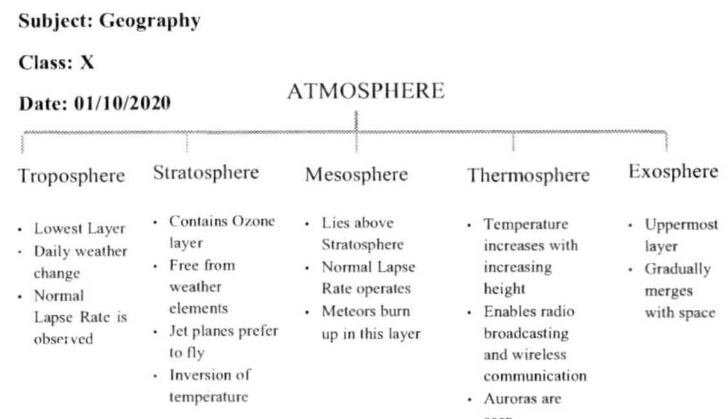

Black Board Work No. 2: The Characteristics of Different Layers

- The trainee-teacher will demonstrate the structure of different layers of the atmosphere based on temperature variation with the help of a chart, namely *"The Levels of Atmosphere"*

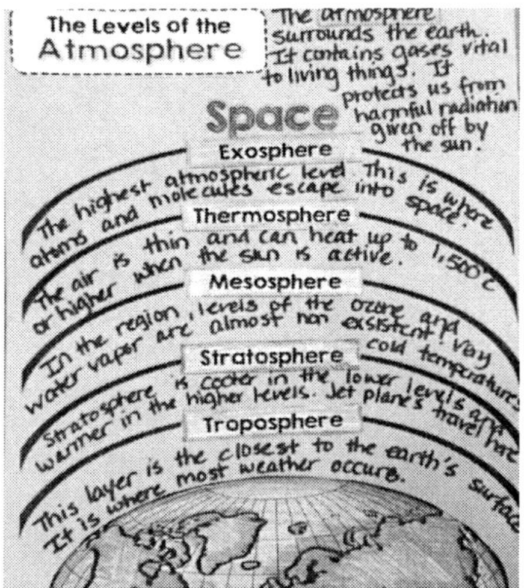

Figure No. 2: Chart Showing the Levels of the Atmosphere

- The trainee teacher will demonstrate some real-life pictures of Auroras in flash- card.

Figure No. 3: Auroras

<table>
<tr><th>Learning Areas</th><th>Related Strategies</th></tr>
<tr><td></td><td>

- The trainee teacher will show some short videos on the burning of meteors in the Mesosphere and Jet planes flying through the Stratosphere. [URL: https://www.youtube.com/watch?v=O4QrnC9vLs8 ; https://www.youtube.com/watch?v=pxuhyQLykao]
- The trainee will also demonstrate that the troposphere is the densest layer by stacking books one above the other.

Figure No. 4: Stacking of Books

- The trainee teacher will interact with the students by asking them some questions:
  - When you visit the mountains, temperature increase or decrease?
  - What will happen to us in the absence of the ozone layer?
  - As student how can you contribute to preventing ozone depletion?
  - In which layer do you think communication satellites are located?
  - Why do you think Jet planes prefer to fly through Stratosphere?

</td></tr>
</table>

USE OF TEACHING AIDS

The trainee-teacher will use the following aids to execute the lesson-

Type of Teaching Aid	*Specification of Teaching Aids*	*Broad Concept*	*Uses of Teaching Aids*
General Aids	Board/Council Prescribed Geography Textbook of Class 10	Detailed information about the composition of the atmosphere and its various layers	The book helps the students to connect the information in the textbook with the teacher's teaching
	Blackboard, Chalk, and Duster	• The Structure of atmosphere • The characteristics of different layers of the atmosphere	• Blackboard work is done to discuss the complex structure of the atmosphere • To make the students attentive the teacher writes down the important characteristics of each layer of atmosphere on the blackboard

Special Aids	Chart	• Layers of Atmosphere Based on Chemical Composition • The Levels of the Atmosphere	• The chart namely "Layers of Atmosphere Based on Chemical Composition" can help students to enhance their understanding of the chemical configuration of the atmosphere • The chart namely "The Levels of Atmosphere" helps to show the layers of the atmosphere based on temperature variation
	Real Images in Flash-card	Auroras	The teacher shows the real image of Auroras to demonstrate the distinctive characteristics of 'Ionosphere'
	Audio-Visual Aids	A short video is shown about the burning of meteors in the earth's Mesosphere and Jet planes flying through the Stratosphere	On this regard, to give exposure to the unique characteristics of the Stratosphere and Mesosphere two short videos can be shown by the teacher.

Probing Questions

i. Why thickness of the troposphere varies from the Equator to the poles?

Ans- The Troposphere is thicker over the equator than the poles because the equator is warmer. Heat differences on the planet's surface cause convection currents to flow from the Equator to the Poles. This implies that the warmer the weather, the thicker the troposphere.

ii. Why in the absence of an ozone layer the biosphere will turn into a blast furnace?

Ans- In the absence of an ozone layer all the UV rays of the sun will reach the earth's surface directly and consequently the temperature of the atmosphere will rise to such an extent that the biosphere will turn into a blast furnace.

Worksheet

A. Identify the correct answer:

i. The layer of the atmosphere where changes in daily weather take place is called: (A)
(a) Troposphere (b) Stratosphere (c) Exosphere (d) Ionosphere

ii. The upper limit of the Mesosphere is called: (R)
(a) Tropopause (b) Stratopause (c) Ozonosphere (d) Mesopause

iii. The densest layer in the atmosphere is: (U)
(a) Exosphere (b) Homosphere (c) Mesosphere (d) Troposphere

iv. Ozone protects mankind from: (E)
(a) UV rays (b) Infrared rays (c) X-rays (d) Gamma rays

v. The country that witnesses polar lights: (A)
(a) Spain (b) India (c) Norway (d) Mexico

B. Give a one-word answer:

i. The upper limit of the Troposphere that separates it from the Stratosphere. (An)
ii. Harmful rays that lead to skin cancer in human beings. (U)
iii. The Coldest layer of the atmosphere. (R)
iv. The continent over which a large ozone hole has developed. (A)
v. The layer that shows heterogeneous character. (R)

C. Fill in the blanks:

i. The decrease of temperature at the rate of 6.5°C/km is known as the ________. (R)
ii. The________ layer absorbs the harmful rays of the sun. (R)
iii. Jet planes prefer to fly through ________ over Troposphere. (An)
iv. Communication satellites are located in ________ layer of atmosphere. (A)
v. Up to the height of ________ km, the composition of atmospheric gasses remains almost the same. (R)

D. Match column A with column B:

Column A	Column B
(i) Troposphere (R)	(a) Meteors burn up.
(ii) Northern polar lights (An)	(b) Aurora Borealis.
(iii) Mesosphere (U)	(c) Presence of Ozone layer.
(iv) Stratosphere (U)	(d) Enables radio broadcasting.
(v) Thermosphere (An)	(e) Lower-most layer of the atmosphere.

[Note: R- Remembering, U-Understanding, A- Applying, An-Analysing, E-Evaluating, C- Creating]

BLUE-PRINT FOR CRITERION REFERENCE TEST ITEMS

Subject: Geography
Class: X
Unit: Atmosphere
Full Marks: 30
Time: 40 minutes
Chosen Sub-unit: Structure of Atmosphere

	Cognitive Process Dimensions									
	R					*U*				
CONCEPT	*OT*	*VST*	*ST*	*LT*	*CT*	*OT*	*VST*	*ST*	*LT*	*CT*
CONCEPT 1	1(1)	1(1)	-	-	-	1(1)	-	1(2)	-	-
CONCEPT 2	3(1)	2(1)	-	-	-	1(1)	1(1)	-	-	-
TM	4	3	0	0	0	2	1	2	0	0
%	13.2%	10%	0%	0%	0%	6.7%	3.3%	6.7%	0%	0%

	Cognitive Process Dimensions									
	A					*An*				
CONCEPT	*OT*	*VST*	*ST*	*LT*	*CT*	*OT*	*VST*	*ST*	*LT*	*CT*
CONCEPT 1	-	-	-	-	-	-	-	-	-	-
CONCEPT 2	2(1)	1(1)	-	-	-	2(1)	-	1(2)	1(3)	-
TM	2	1	0	0	0	2	0	2	3	0
%	6.7%	3.3%	0%	0%	0%	6.7%	0%	6.7%	10%	0%

	Cognitive Process Dimensions											
	E					*C*						
CONCEPT	*OT*	*VST*	*ST*	*LT*	*CT*	*OT*	*VST*	*ST*	*LT*	*CT*	*FM*	*%*
CONCEPT 1	-	-	-	-	-	-	-	-	-	-	05	16.7%
CONCEPT 2	-	-	-	1(3)	-	-	-	-	-	1(5)	25	83.3%
TM	0	0	0	3	0	0	0	0	-	5	30	100%
%	0%	0%	0%	10%	0	0%	0%	0%	0%	16.7%	100%	100%

Key: Cognitive Process Dimension				*Key: Type of Questions*			
R	Remembering	An	Analyzing	OT	Objective type	LT	Long type
U	understanding	E	Evaluating	VST	Very short type	CT	Constructive type
A	Applying	C	Creating	ST	Short type	-	-

1(1) ⟶ 1 Question of 1 mark

TEST ITEMS

Date: 21/03/2024 **Class:** X

Time: 40 minutes **Full Marks**=30

Instruction to the Examinees-

- Read the questions carefully before answering them.
- Extra time will be given for reading the questions.
- Do not copy the questions.
- Attempt all the questions.

I) Fill in the Blanks (OT) (5×1=5)

a) is the layer in the atmosphere that shows heterogeneous character. *(Remembering, concept 1)*

b) Up to the height of 100 km, the composition of atmospheric gasses remains almost the ________________ *(Understanding, concept 1)*

c) ________________ is the layer of the atmosphere where changes in daily weather takes place. *(Applying, concept 2)*

d) Jet planes prefer to fly through ________ over Troposphere. *(Analysing, concept 2)*

e) Ozone protects mankind from______________ rays. *(Remembering, concept 2)*

II) State True or False (OT) (5×1=5)

a) Thickness of the Troposphere remains constant around the Equator and near the poles. *(Understanding, concept 2)*

b) Exosphere is the lowermost layer of the atmosphere. *(Remembering, concept 2)*

c) Normal Lapse rate operates in Mesosphere. *(Remembering, concept 2)*

d) Temperature increases with increasing height in the Stratosphere due to the presence of the Ozone layer. *(Analysing, concept 2)*

e) Air-conditioners are major household sources of generating CFCs. *(Applying, concept 2)*

III) Answer the Following in One or Two Words (VST) (5×1=5)

a) State the extension of the Homosphere. *(Remembering, concept 1)*

b) Name the upper limit of the Troposphere. *(Remembering, concept 2)*

c) Name the continent over which a large ozone hole has developed. *(Applying, concept 2)*

d) Name the layer where meteors slow down and burn up. *(Understanding, concept 2)*

e) Give another name for the northern polar lights. *(Remembering, concept 2)*

IV) Write in One or Two Sentences (ST) (2×2=4)

a) Differentiate between Homosphere and Heterosphere? *(Understanding, concept 1)*

b) Give a reason for atmospheric pressure being lowest in Exosphere. *(Analysing, concept 2)*

V) Answer the Following in Brief (LT) (2×3=6)

a) Compare and contrast between the Troposphere and Stratosphere. *(Analysing, concept 2)*

b) Validate the need or importance of the presence of the Thermosphere in the atmosphere and its impact on human lives. *(Evaluating, concept 2)*

VI) Construct a Diagram on the Following (CT) (1×5=5)

a) Create a diagram showing the different layers of the atmosphere based on temperature variation and label it properly in accordance to their extension and significance. *(Creating, concept 2)*

ITEM-WISE ANALYSIS

Item No.	*Concept*	*Type of Question*	*Instructional Objectives*						*Marks Allotted*	*Approximate Difficult Level*	*Proposed Time*
			R	*U*	*A*	*AN*	*E*	*C*			
I(a)	1	OT	✓						1	E	0.5
I(b)	1	OT		✓					1	M	1
I(c)	2	OT			✓				1	M	1
I(d)	2	OT				✓			1	M	1
I(e)	2	OT	✓						1	E	0.5
II(a)	2	OT		✓					1	M	1
II(b)	2	OT	✓						1	E	0.5
II(c)	2	OT	✓						1	M	1
II(d)	2	OT				✓			1	D	1.5
II(e)	2	OT			✓				1	D	1.5
III(a)	1	VST	✓						1	M	1.5
III(b)	2	VST	✓						1	E	0.5
III(c)	2	VST			✓				1	M	1
III(d)	2	VST		✓					1	M	1
III(e)	2	VST	✓						1	M	1
IV(a)	1	ST		✓					2	M	3.5
IV(b)	2	ST				✓			2	D	4
V(a)	2	LT				✓			3	M	5
V(b)	2	LT					✓		3	D	6
VI(a)	2	CT						✓	5	D	7
Total									30		40 mins

E- Easy, M- Medium, D-Difficult

Key: Cognitive Process Dimension				*Key: Difficulty Level*	
R	Remembering	AN	Analyzing	E	Easy
U	Understanding	E	Evaluating	M	Medium
A	Applying	C	Creating	D	Difficult

Distribution of Marks

Serial No. of Question	*Marks*
I	5
II	5
III	5
IV	4
V	6
VI	5
Total	**30**

GUIDELINES FOR COURSE OUTCOME (CO)

Semester	XX
Paper Code	XXX
Paper Title	XXX XXX
No. of Credits	X
Theory/Practicum	Theory
No of periods assigned	X teaching hours in accordance to the period duration & content
Course descriptions/	Type of paper- general/ Special/ Method
Outcome	Mention the 4/5 broad objectives of the content in the paper.

GUIDELINES FOR PROGRAMME OUTCOME (PO)

1. To foster an effective integration of diverse knowledge domains, aligning with the essential training requirements for prospective educators.
2. To ensure the curriculum remains contemporary, professional, and imbued with practical, skills-oriented learning experiences.
3. To equip trainee-teachers with the competencies necessary to address academic, societal, and national educational demands.
4. To cultivate a positive life outlook and nurture holistic personality development among trainee-teachers, fostering a commitment to community service.
5. To instill resilience and adaptability in trainee-teachers to confront the multifaceted challenges of the modern age.
6. To instill discipline and foster the cultivation of moral values among trainee-teachers.
7. To motivate and empower trainee-teachers to inspire and mentor future generations.
8. To incorporate opportunities for trainee-teachers to engage with innovative teaching methodologies and emerging educational technologies, preparing them for dynamic educational landscapes.

A Final Summarised Format of the Item Analysis

Name of the Institution

Data Sheet for Mapping Course and Programme Outcomes of the Question Paper

SESSION:			*PROGRAMME CODE:*		
COURSE CODE:			*SEMESTER:*		
DURATION:			*FULL MARKS:*		
Question Number	*Allotted Marks*	*Type of Question*	*Difficulty Level*	*Course Outcome (CO)*	*Programme Outcome (PO)*

Signature of Co-ordinator (with date) **Signature of Moderator (with date)**

*Type: MCQ/ Descriptive/ Objective/ Problem Solving/ Analytical

** Difficulty: Basic/ Moderate/ High

SCORING KEY

I. a) Heterosphere
   b) Same
   c) Troposphere
   d) Stratosphere
   e) UV

II. a) False
   b) False
   c) True
   d) True
   e) True

III. a) From surface up to 100 km
   b) Tropopause
   c) Antarctica
   d) Mesosphere
   e) Aurora Borealis

IV. a) The differences between Homosphere and Heterosphere are as follows:

Homosphere	*Heterosphere*
1. Extends from the surface up to a height of 100 km.	1. Extends from 100 to up to 10,000 km above the surface of the earth.
2. The composition and proportion of atmospheric gasses remain constant.	2. The percentage of gasses varies and shows heterogeneous character.

IV. b) Exosphere being the uppermost layer of the atmosphere has air that is extremely thin and rarefied. Hence, this makes the atmospheric pressure the lowest in this layer.

V. a) Troposphere and stratosphere layer of the atmosphere show some different as well as some similar characteristics:

Similarities:

i) In both Troposphere and Stratosphere atmospheric pressure decreases with increasing height.

Differences:

i) All dust particles and water vapour are present in the Troposphere, therefore all changes in daily weather conditions take place in this layer. On the other hand, absence of water vapour in the Stratosphere makes this layer free from weather disturbances.

ii) Temperature decreases with altitude in the Troposphere following the Normal Lapse rate. Whereas in the Stratosphere temperature increases with increasing height.

V. b) From the point of view of the present world scenario, the need or importance of the Thermosphere is immense in our lives:

- Electrically charged particles of ions and electrons are found here, so radio waves coming from the earth's surface are reflected by the earth, so it enables radio broadcasting and wireless communication across the world.
- Polar lights called the Aurora Borealis and Aurora Australis of the Northern and Southern hemispheres respectively are developed in this layer which acts as an important source of lighting during the months of darkness in the poles.

VI. a)

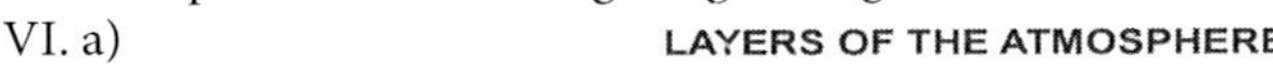

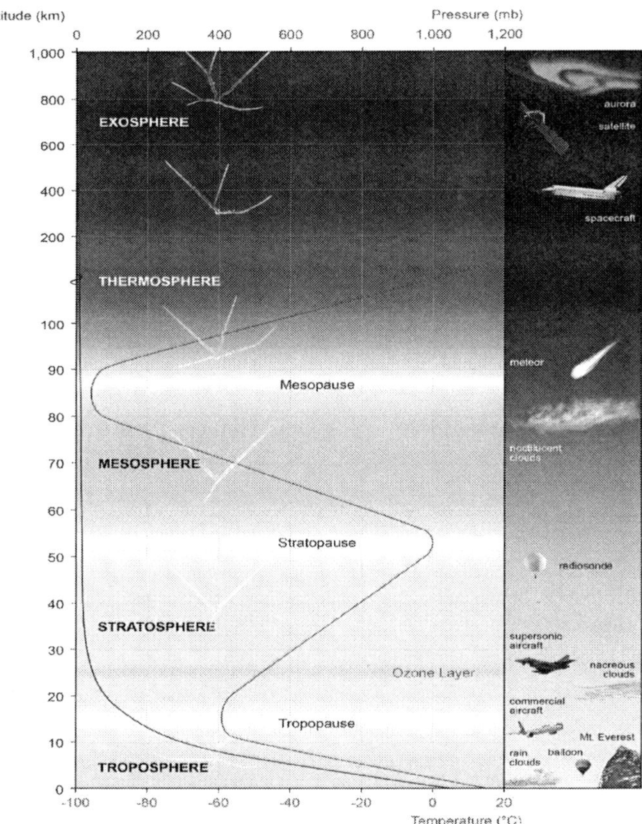

Diagram showing the Different Layers of the Atmosphere based on Temperature Variation

18

Definition of Achievement Tests: Geography

Introduction

The progress of a student can be measured on the basis of an achievement test, and a number of educational and vocational decisions on their performance can be taken by a teacher in a classroom environment. Since achievement tests help in the instructional development of an individual, it is necessary that a teacher be deft in various aspects of it.

An achievement test is defined as 'a systematic procedure for determining the amount a student has learned through instruction' (Gronlund, 1977). It is a test of performance to measure the knowledge and skill of a student on a particular content area compared with the performance of a general reference group of students on the same. However, the individual student's attainment is time-specific, and it may change under different circumstances. According to Thorndike and Hagen (1969), 'The type of ability test that describes what a person has learned to do is called an achievement test'.

Purpose of Achievement Tests

Achievement tests have practical utility for teachers and administrators of academic institutions. On the basis of it a teacher can find out the general range of capabilities of students in a class, and accordingly, one may select appropriate materials of instructions so that maximum instructional benefit can be provided to all individuals. The test is also helpful in the categorization of the students, working at their best capacity, as either weak or advanced, and necessary measures can be taken up separately to enhance the progress of both.

For the administrators, achievement test help in classifying school objectives; improving curriculum; preparing for measures to be taken separately for the weak and advanced students; providing incentives to the students based on their performances; and determining the overall educational attainment of an institution, etc. The management may take appropriate steps in light of the achievements for the betterment of it.

A test is defined as a standardized situation designed to elicit a sample of an individual's behaviour. Tests are classified as 1. Test of ability 2. Test of typical performance and 3. Educational achievement test.

Types of Evaluation or Tests

Achievement tests are practically a type of evaluation. Following are the different types of achievement tests (Mangal and Mangal, 2008):

1. ***Standardized tests (usually written with objective type of questions)***
   (a) General achievement (i) Test Battery (ii) Specialized
   (b) Diagnostic

2. ***Teacher-made informal tests***
   (a) Written paper test (i)Short answer types (ii) Very short answer types (iii) Objective types
   (b) Oral test
   (c) Practical test
   Tests are categorized as: (1) Formative, e.g., Unit test; (2) Summative, e.g., Half-yearly/Annual examinations (3) Diagnostic, e.g., Remedial test and (4) Prognostic, e.g., career counseling.

Essential Qualities of A Good Achievement Test

A good achievement test should be based on all difficulty levels and it should include a sufficient number of test items to cover all aspects of the topic. The questions should be unambiguous and properly framed to seek different knowledge and skill according to the suitable age groups. The test should be as objective as possible, and it should be standardized so that it can be compared to any other test of a similar type. The pupils should not feel bored or fatigued while going through the test. A good achievement test always encourages students to become creative and self-reliant. This also should be easy to administer and score.

Essential Points for Construction of Test Items

The following points are to be adhered to in developing balanced and objective-based questions:

1. Instructional objectives 2. Content analysis 3. Form of question 4. Language of questions
5. Directional words 6. Difficulty level 7. Marking schemes.

Stages of Planning for an Achievement Test in Geography

The following are the important steps normally adopted in planning for an achievement test:

(i) *Determination of the purpose* - There should be a specific purpose of an achievement test, e.g., to find out and categorize a group of students, say of Class VIII, into advanced learners, average learners, or slow learners according to their performance in the test.
(ii) *Selection of the target group* - The target group is selected on the basis of purpose and it should be clearly stated. Usually, the target groups may include either a class or a certain group of students of similar age.
(iii) *Selection of the topic* - The topic or sub-topic may be selected on the basis of availability of time, weightage of the test (i.e., marks assigned), and other resources affecting the test.
(iv) *Analysis of the content* - The content of the test should be chosen purposively from a suitable area of the subject of a particular class and then it is to be unitized and sub-unitized according to the specific conditions.
(v) *Development of the instructional objectives* - A group of cognitive psychologists, curriculum theorists and instructional researchers, and testing and assessment specialists published in 2001 a revision of Bloom's Taxonomy with the title *A Taxonomy for Teaching, Learning, and Assessment* to examine the educational objectives of the learners.

(vi) *Preparation of the table of specifications or blueprint* - This includes a detailed question-wise distribution of marks over specific objectives, topics, and forms of questions. Thus, all important factors like objectives to be tested, contents to be covered, types of questions to be asked and total time to be given should be kept in mind while preparing this table of specifications.

Table of Specifications Based On Criterion-Referenced Test (Crt)

Main topic / Unit :	*Essay Type*						*Short Answer Type*						*Very Short Answer Type*						*Objective Type*						*Total marks*	*% of marks*
Sub-topics/ Sub-units:	R	U	A	A	E	C	R	U	A	A	E	C	R	U	A	A	E	C	R	U	A	A	E	C		

Note: R = Remembering, U = Understanding, A = Applying, A = Analyzing, E = Evaluating, C = Creating

(vii) *Selection of item format on the basis of the table of specifications* - This can be on the basis of either weightage to objectives or weightage to forms of questions or sometimes on the basis of weightage to contents. This table of specifications should give self-confidence about the justice of the test. However, usually, it is done according to the table of specifications on the basis of weightage to forms of questions. This is shown in the following table.

(viii) *Development of the test items (i.e., question paper) as per the table of specifications* - Questions to be constructed as per ***(vi)*** (Table of specifications) and ***(vii)*** (Item format). However, the following points may help in preparing the test format (e.g., a question paper) :

- (a) Long and short items should be kept in separate sections with specific instructions against each of them.
- (b) The test items should be arranged in order of ascending difficulty level; that means it should proceed from easy to more difficult ones.
- (c) For the objective items the multiple-choice format is the most suitable one because of its reliability and validity. Other types may also be chosen, but they should be similar to prevent loss of time in reading the instructions.
- (d) It is better to frame a few extra items so that at the final review the unsuitable ones can be excluded.

(ix) *Preparation of the answer key* - Appropriate answer key should be prepared to compare the main points expected from the students. This is also necessary for proper evaluation so that a group of examiners can evaluate following the same set of the answer key, which will ultimately minimize any kind of biasness and personal error of the evaluators. The answer key may seek to incorporate the following points:

- (a) Proper division of marks according to sections and sub-sections especially for the essay-type questions
- (b) The points or steps expected in the answer

Item Format on the Basis of the Table of Specifications (TOS)

Class: Full Marks:
Subject: Topic: Time:

Unit	*Remembering*				*Understanding*				*Applying*				*Analyzing*				*Evaluating*				*Creating*				*% of Marks*
Sub-unit	*ET*	*SAT*	*VSAT*	*OT*	*ET*	*SAT*	*VSAT*	*OT*	*ET*	*SAT*	*VSAT*	*OT*	*ET*	*SAT*	*VSAT*	*OT*	*ET*	*SAT*	*VSAT*	*OT*	*ET*	*SAT*	*VSAT*	*OT*	
Marks assigned																									
Total marks																									
% of marks																									

Note: ET = Essay Type, SAT = Short Answer Type, VSAT= Very Short Answer Type, OT = Objective Type

(c) Number of lines or words expected for the short and very short type questions
(d) The weight given to the steps and important points to be covered
(e) Requirement of explanations and illustrations and their weightage
(f) Weightage for diagrams, tables, sketches, etc.

(x) Administration of the test on the target group already selected

This stage includes advance announcement of the test with regard to date, time, and venue; and other instructions related to the accessories, e.g., admit card, drawing and writing instruments, charts, models, etc. to be brought into the institution by the students. Sometimes, prior intimation is also given regarding the type of test to be conducted. However, personal problems- either physical or social, or psychological, of the pupils should not flout the test in any manner. Clear instructions should also be given to start or end the test, and also during the test if so required.

(xi) Scoring and appraising - Scoring can be done either manually or mechanically depending on the type of questions and available resources. Appraisal of the test is also called 'item-analysis' which means each item of the test is analyzed to know for its difficulty level or otherwise on the basis of the responses given by the students against the individual items.

(xii) Using the test results - The primary usage of the test results is to decide about the promotion of the pupils to the next higher grade. Besides, the results can also be helpful in enabling a teacher to introspect and reorganize the teaching-learning activities. The measurements/ scores in the Geography of a student in achievement test will indicate the meaningful cognitive, affective and psychomotor outcomes and thereby designate the level of mastery of the subject.

Through achievement tests, teachers can be assured that they are measuring students' learning across a wide range of content and reading as well as cognitive processes requiring higher-order thinking. It helps in assisting the teacher to select the most appropriate teaching strategy for creating real geographical knowledge. Teachers are able to determine what topic(theory/practical) is being stressed and also assist in the preparation of tests that reflect what students have assimilated and also limit the amount of time spent on each unit and also consider that no important objective or content area has been omitted.

Model of an Achievement Test Paper of Geography

In this case, any sample of geography question paper can be taken at School levels (Classes IX, X, XI and XII) affiliated to National and State Boards.

Construction of Achievement Test
Content Analysis

Class: X **Subject: Geography**

Units	*Subunits*
Biosphere	• Biosphere • Ecosystem & Ecology • Trophic Structure • Food chain & Food web • Energy Flow

Units	*Subunits*
Soil	• Genesis of soil • Soil profile • Properties of soil • Formation of soil • Type of soils

Instructional Objectives

Unit: Biosphere

Remembering: Students will be able to

- Define ecosystem. (factual)
- Recall the characteristics of ecosystem. (factual)
- Define food chain and food web. (factual)

Understanding: Students will be able to

- Distinguish between food chain and food web. (Conceptual)
- Discuss the phases of energy flow. (Conceptual)
- Interpret different components of biosphere. (Conceptual)

Applying: Students will be able to

- Identify different types of ecosystem in environment. (Procedural)
- Determine the role of different organism found in surroundings in energy flow. (Conceptual)

Analyzing: Students will be able to

- Classify ecosystem. (Conceptual)
- Analyze the trophic structure of ecosystem. (Conceptual)

Evaluating: Students will be able to

- Evaluate the role of different species with respect of ecological balance. (Meta-cognition)
- Justify the interdependence of different species of environment. (Meta-cognition)
- Justify the importance of ecological conservation. (Meta-cognition)

Creating: Students will be able to

- Draw the diagram of energy flow. (Procedural)
- Draw the diagrams of food chain and food web. (Procedural)

Unit: Soil

Remembering: Students will be able to

- Define soil. (factual)
- Recall the equation of soil forming factors. (factual)

Understanding: Students will be able to

- Distinguish between eluviation and illuviation. (Conceptual)
- Distinguish between pedocal and pedalfar. (Conceptual)
- Describe soil forming factors. (Conceptual)

Applying: Students will be able to

- Identify different soils of India. (Conceptual)
- Identify different soils suitable for cultivating different crops. (Conceptual)

Analyzing: Students will be able to

- Classify soil. (Conceptual)
- Divide soil forming factors. (Procedural)
- Analyze zonal, inter-zonal and azonal soils. (Conceptual)

Evaluating: Students will be able to

- Examine the role of soil in human life. (Meta-cognition)
- Evaluate the role of soil in environment. (Meta-cognition)

Creating: Students will be able to

- Draw the diagram of soil profile. (Procedural)
- Draw maps for showing different zonal soils in the earth. (Procedural)

Unit Wise Distribution of Marks

Units	*Item Number*	*Marks*	*% of Marks*
Biosphere	1A.i, ii, 1B.i, ii, 1C iii, iv, 1D i,ii,2 i, ii, 3 ii, 4	26	52
Soil	1A.iii, iv, 1B.iii, iv, 1C i, ii, 1D iii, iv, 2iii, iv, 3 i, iii, iv	24	48
Total		**50**	**100**

Unit Wise Distribution of Marks

Objectives Wise Distribution of Marks

Objectives	*Item Number*	*Marks*	*% of Marks*
Remembering	1A.i,1B.i,ii,1C iv,1D i	5	10
Understanding	1A.ii,iii, 1B.iv, 1C ii, 1D ii,3i, 4	19	38
Applying	1Aiv, 1Biii, 1Ci, iii,1Diii	5	10
Analyzing	1D iv,2i,3iv	7	14
Evaluating	2iii,3iii	6	12
Creating	2ii,iv, 3ii	8	16
Total		**50**	**100**

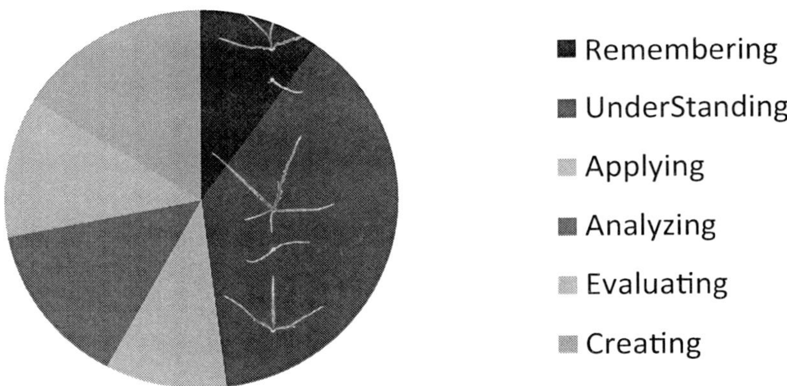

Type of Question Wise Distribution of Marks

Type of Questions	*Item Number*	*Marks*	*% of Marks*
Objective	1A.i,ii,iii,iv, 1B.i,ii,iii,iv,1C i,ii,iii iv,1D i,ii,iii,iv	16	32
Very Short	2i,ii,iii,iv	8	16
Short	3i,ii,iii,iv	16	32
Essay	4	10	20
Total		50	100

Type of Question Wise Distribution of Marks

Table of Specification

Objectives	*Remembering*				*Understanding*				*Applying*				*Analyzing*				*Evaluating*				*Creating*					
Units	*O*	*VS*	*S*	*E*	*O*	*VS*	*S*	*E*	*O*	*VS*	*S*	*E*	*O*	*VS*	*S*	*E*	*O*	*VS*	*S*	*E*	*O*	*VS*	*S*	*E*	*TM*	%
Biosphere	1A.i (1) 1B.i,ii(2) 1Civ(1) 1Di(1)				*1A.ii(1) 1Dii(1)			4 (10)	1Ciii(1)					2i (2)								2ii(2)	3ii(4)		26	52
Soil					1A.iii(1) 1B.iv(1) 1Cii(1)		3i (4)		1Aiv(1) 1B.iii(1) 1Ci(1) 1Diii(1)				1Div (1)		3iv (4)			2iii (2)	3iii (4)			2iv(2)			24	48
	5				5		4	10	5				1	2	4			2	4			4	4			
Total Marks	5				19				5				7				6				8				50	
% of Marks	10				38				10				14				12				16				100	

* Outside the bracket question or item no. and inside the bracket marks O= Objective Type VS: Very Short Type S: Short Type E: Essay Type

Name of the School

Achievement Test

Class: X **Full Marks: 50**
Subject: Geography **Time: 2 hours.**

Instruction:
i. Students are instructed to write in your own words.
ii. All questions are compulsory.

1. A Fill in the blanks. **1×4=4**

i. _______ coined the term 'Ecosystem' in 1935.
ii. Upper the trophic level _______ the number of consumer.
iii. The influence of parent rock causes _______ soil.
iv. In Birbhum, the high concentration of iron helps to form _______ soil.

1. B Choose the appropriate answer. **1×4=4**

i. Ecosystem is the system of –
   a) Open System
   b) Close System
   c) Quasi-open System
ii. According to 10% law, at each step of ecosystem the actual energy
   a) Increases
   b) Decreases
   c) Remains same
iii. The alluvial soil at the side of Hooghly river bank is a type of-
   a) Zonal Soil
   b) Azonal Soil
   c) Inter-zonal Soil
iv. The regolith is formed due to-
   a) Weathering
   b) Humification
   c) Capillary action

1. C Select true or false statements. **1×4=4**

i. Rendzina soil is found at the Karst region.
ii. The minerals are deposited in the B layer of soil due to illuviation.
iii. The grazing ecosystem is found in oceanic environment.
iv. The energy flow in food chain is multidirectional.

1. D Give answer with one or two sentences. **1×4=4**

i. What is biomass?
ii. Give on difference between producer and consumer.
iii. Where is found sierozem soil in India?
iv. How many minerals are there to form white and gray colour in soils.

2. Give answer within 50 words. **2×4=8**

i. Classify the components of ecosystem.
ii. Draw a flow chart to illustrate trophic levels in ecosystem.
iii. The soil formation takes less time upon permeable rocks- Justify the reason.
iv. Give a diagram of soil catena.

3. Give answer within 100 words. **4×4=16**

i. Differentiate between pedocal and pedalfer soils.
ii. Give a layout of pond ecosystem with a diagram.
iii. Justify the reasons of formation of podzol soils in wet-temperate region.
iv. Analyze the cross-section of a soil profile.

4. Give answer within 200 words. **10×1=10**

Describe the phases of energy flow in ecosystem.

Answer Key

1A i. Tansley
1A ii. Lower
1A iii. Endomorphic
1A iv. red
1B i. a) open system
1B ii.b) Decreases
1B iii.b) Azonal Soil
1B iv.a) Weathering
1C i. True
1C ii. True
1C iii. False
1C iv. False
1D i. Biomass is the weight or total quantity of living organisms commonly referred to a unit area or volume of habitat.
1D ii. The producers generate food for themselves and consumers do not produce anything, instead eating producers, other consumers or both.
1D iii. The sierozem soil in India is found at Thar desert.
1D iv. Mineral sault, limestone and quartz are there to form white and grey colour in soils.
2i.

Components of Ecosystem

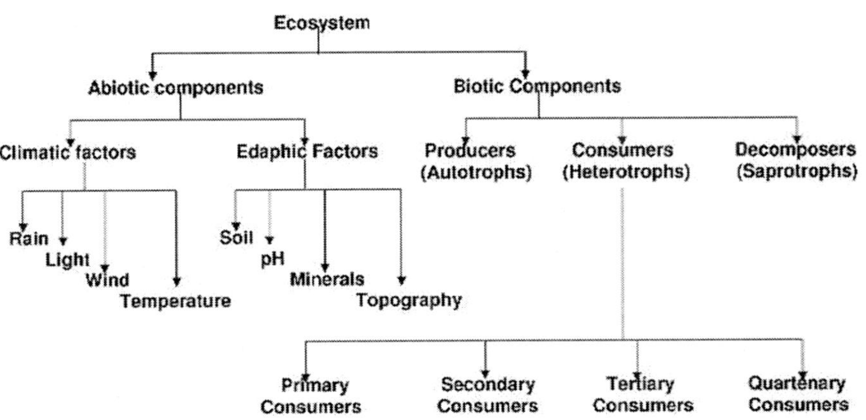

2ii.

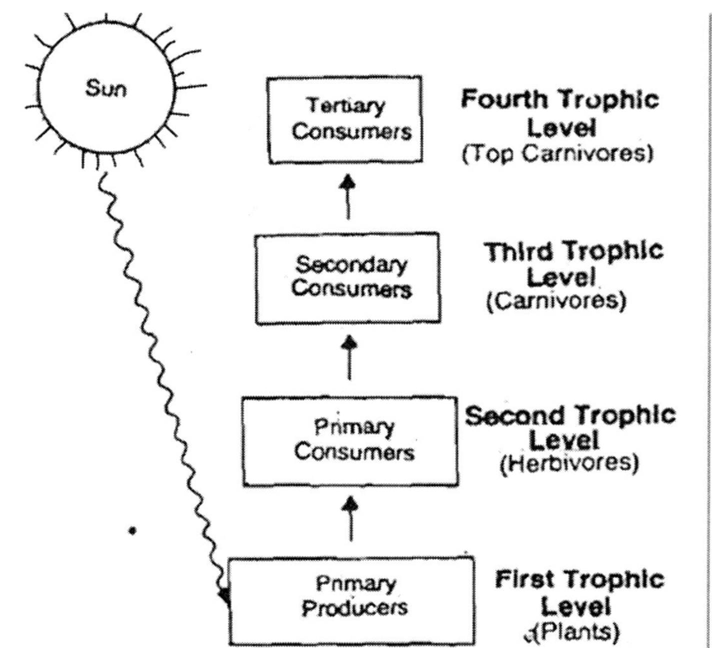
Sun
Tertiary Consumers
Fourth Trophic Level
(Top Carnivores)
Secondary Consumers
Third Trophic Level
(Carnivores)
Primary Consumers
Second Trophic Level
(Herbivores)
Primary Producers
First Trophic Level
(Plants)

2iv.

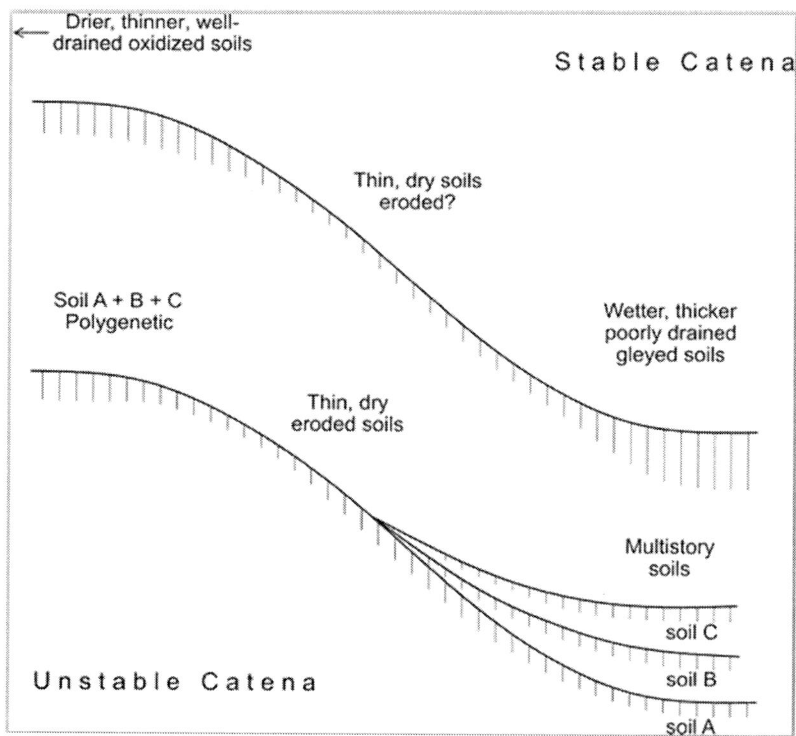
Drier, thinner, well-drained oxidized soils
Stable Catena
Thin, dry soils eroded?
Soil A + B + C Polygenetic
Wetter, thicker poorly drained gleyed soils
Thin, dry eroded soils
Multistory soils
soil C
soil B
soil A
Unstable Catena

3i. **Pedocal**

- Dry, semi arid climate
- Little organic matter
- No mineral leaching
- High limestone content
- parties

Pedalfer

- Enriched with aluminum and iron
- Greater organic matter and leaching
- Found in areas with high temp. and lots of forest cover

3ii.

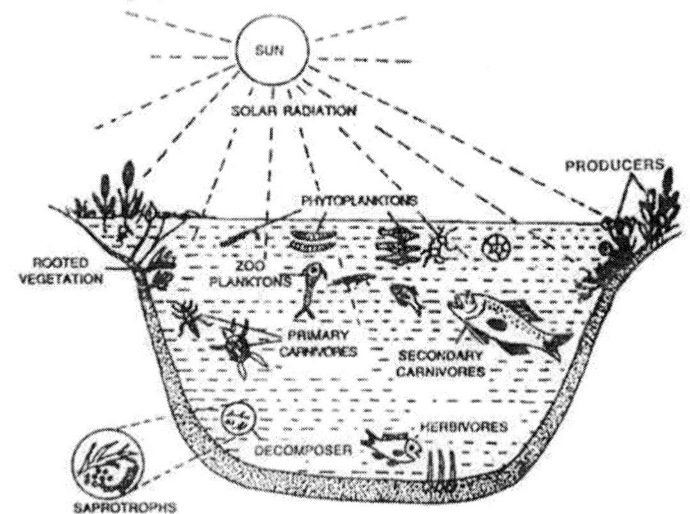

PONDS AS ECO-SYSTEM

Pond Ecosystem falls in fresh water ecosystem which is further divided into three parts which are littoral zone, open water zone and deep water zone. The littoral zone is a zone which is an area near the shore where sunlight follows it way down and allows the plants to grow. Open water zone which actually is well known as pelagic zone. It is that zone where water is neither close to the bottom nor near the shore. The last is the deep water zone which is the lowest level lying above the sea and below the thermocline. These three types can be understood deeply. Some differentiate Pond Ecosystem in other way. It is classified in five types which are salt ponds, garden ponds, freshwater ponds, vernal ponds and underground ponds. Salt ponds are one which are near the sea side and consist brackish water. Garden ponds are those which are artificial in nature. Freshwater pool can form anywhere because of rains. Vernal ponds are those which are created because of depression in the ground and are seasonal. Lastly underground ponds that are created near the rocks.

3.iv A soil horizon is a layer parallel to the soil surface whose physical, chemical and biological characteristics differ from the layers above and beneath. Horizons are defined in many cases by obvious physical features, mainly colour and texture.

O) Organic surface layer: Plant litter layer, the upper part often relatively undecomposed, but the lower part may be strongly humified.

A) Surface soil: Layer of mineral soil with most organic matter accumulation and soil life. Additionally, due to weathering, oxides (mainly iron oxides) and clay minerals are formed and accumulated. It has a pronounced soil structure. But in some soils, clay minerals, iron, aluminum, organic compounds, and other constituents are soluble and move downwards. When this eluviation is pronounced, a lighter coloured E subsurface soil horizon is apparent at the base of the A horizon. The A horizon may also be the result of a combination of soil bioturbation and surface processes that winnow fine particles from biologically mounded topsoil. In this case, the A horizon is regarded as a "biomantle".

B) Subsoil: This layer has normally less organic matter than the A horizon, so its colour is mainly derived from iron oxides. Iron oxides and clay minerals accumulate as a result of weathering. In soil, where substances move down from the topsoil, this is the layer where they accumulate. The process of accumulation of clay minerals, iron, aluminum, and organic compounds, is referred to as illuviation. The B horizon has generally a soil structure.

C) Substratum: Layer of non-indurated poorly weathered or unweathered rocks. This layer may accumulate more soluble compounds like $CaCO_3$. Soils formed *in situ* from non-indurated material exhibit similarities to this C layer.

R) Bedrock: R horizons denote the layer of partially weathered or unweathered bedrock at the base of the soil profile. Unlike the above layers, R horizons largely comprise continuous masses (as opposed to boulders) of hard rock that cannot be excavated by hand. Soils formed *in situ* from bedrock will exhibit strong similarities to this bedrock layer.

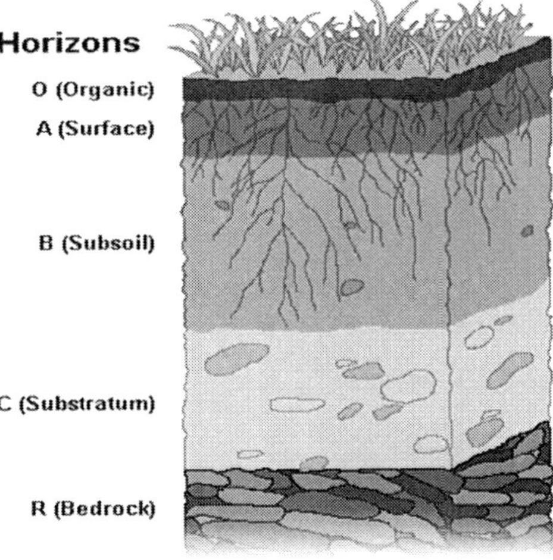

4. The radiant energy is in the form of electromagnetic waves which are released from the sun during the transmutation of hydrogen to helium. The chemical energy stored in the food of living organisms is converted into potential energy by the arrangement of the constituent atoms of food in a particular manner. In any ecosystem there should be unidirectional flow of energy. This energy flow is based on two important Laws of Thermodynamics which are as follows:

   (1) *The first law of Thermodynamics:* It states that the amount of energy in the universe is constant. It may change from one form to another, but it can neither be created nor destroyed. Light energy can be neither created nor destroyed as it passes through the atmosphere. It may, however, be transformed into another type of energy, such as chemical energy or heat energy. These forms of energy cannot be transformed into electromagnetic radiation.

(2) *The second law of Thermodynamics:* It states that non-random energy (mechanical, chemical, radiant energy) cannot be changed without some degradation into heat energy. The change of energy from one form to another takes place in such a way that a part of energy assumes waste form (heat energy). In this way, after transformation the capacity of energy to perform work is decreased. Thus, energy flows from higher to lower level.

Main source of energy is sun. Approximately 57% of sun energy is absorbed in the atmosphere and scattered in the space. Some 35% is spent to heat water and land areas and to evaporate water. Of the approximately 8% of light energy striking plant surface, 10% to 15% is reflected, 5% is transmitted and 80 to 85% is absorbed; and an average of only 2% (0.5 to 3.5%) of the total light energy striking on a leaf is used in photosynthesis and rest is transformed into heat energy.

Energy flow in Ecosystems:

Living organisms can use energy in two forms radiant and fixed energy. Radiant energy is in the form of electromagnetic waves, such as light. Fixed energy is potential chemical energy bound in various organic substances which can be broken down in order to release their energy content.

Organisms that can fix radiant energy utilizing inorganic substances to produce organic molecules are called autotrophs. Organisms that cannot obtain energy from abiotic source but depend on energy-rich organic molecules synthesized by autotrophs are called heterotrophs. Those which obtain energy from living organisms are called consumers and those which obtain energy from dead organisms are called decomposers (Figure 3.7).

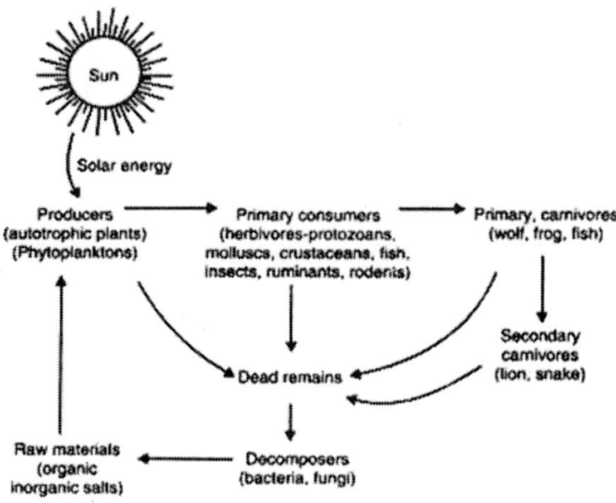

Fig. 3.7. Flow of energy at different levels of ecosystem.

When the light energy falls on the green surfaces of plants, a part of it is transformed into chemical energy which is stored in various organic products in the plants. When the herbivores consume plants as food and convert chemical energy accumulated in plant products into kinetic energy, degradation of energy will occur through its conversion into heat. When herbivores are consumed by carnivores of the first order (secondary consumers) further degradation will occur. Similarly, when primary carnivores are consumed by top carnivores, again energy will be degraded.

19

Simulation Skills: Geography

Skill of Introducing a Lesson

Topic/Unit	*Main Points*	*Components*	*Procedural Details*
Natural Vegetation of India	• Concept and Importance of Natural Vegetation • Types of Natural Vegetation • Measures for Conservation of Forests • Social-Forestry and Agro-Forestry	Securing attention	The teacher will secure the attention of the student by showing a small movie clip demonstrating the forests of Sunderban.
		Assessing motivational level	The teacher will assess the motivational level of the student by asking some questions: ➢ Identify the forest that you have just seen in the movie picture. • Forest of Sunderban ➢ What type of forest is this? • Mangrove forest ➢ Have you seen any other forests either in movies or in pictures or in reality? • Yes ➢ Tell me the name of some other forests that you have seen. • Evergreen forest, Deciduous forest
		Linking with past experience	The teacher will ask questions to relate the previous knowledge of the learners with the unit. For example: ➢ What type of resource is natural vegetation? • Filled resource ➢ Think a little more and then answer. • Flow resource ➢ How does rainfall distribution influence the growth of natural vegetation? • Heavy rainfall, dense vegetation, low rainfall scarcity of vegetation.
		Specifying the points to be taken up for teaching	The teacher will specify the main points to be taken up for teaching: • Definition and importance of natural vegetation. • Types of natural vegetation found in India. • Characteristic features of different types of natural vegetation. • Measures for conservation of natural vegetation. • Concept of social forestry and agro-forestry.
		Using an appropriate device	The teacher can use an appropriate device such as: • A small clipping from the documentary Sunderban: It's not about the tiger only' to attract the attention of the student.

Skill of Questioning

Topic/Unit	*Main Points*	*Components*	*Procedural Details*
Natural Vegetation of India	• Concept and Importance of Natural Vegetation • Types of Natural Vegetation • Measures for Conservation of Forests • Social-Forestry and Agro-Forestry	Precision and clarity of language	The teacher will use precise, clear, relevant, grammatically correct, and accurate language while questioning such as: • What is the importance of natural vegetation? • Why tropical evergreen forests are found on the western slope of the Western Ghats?
		Link with specific objectives	The teacher should ask questions linked with specific objectives. The questions should be like this: ➢ Name some important species of trees found in mountain forests (Remembering). ➢ What is natural vegetation? • The vegetation that grows under natural conditions. ➢ What is the difference between natural vegetation and the vegetation that we grow in our house? • The one that we grow in our house requires care and maintenance but natural vegetation does not require so. ➢ Why do littoral forests develop pneumatophores? (Understanding) ➢ Why conservation of forests is important? (Applying)
		Using the student's response for further questioning	The teacher uses the student's response for further questioning, such as: ➢ Where are tropical evergreen forests found? • Western slopes of the Western Ghats. ➢ Why are they found there? • Because the region receives more than 200 cm of rainfall. ➢ Why does the region receive more than 200cm of rainfall? • Because the moisture-laden wind of the Arabian sea branch of monsoon when ascends along the slopes of the Western Ghats cools and condenses and brings heavy rainfall on the western slopes of the Western Ghats.
		Prompting	The teacher prompts answers to the students in case they can't answer such as: ➢ Why coniferous trees are conical in structure? • The students remain silent ➢ If you place a ball on the top of a mountain will the ball remain there or it will slide down? • The ball will slide down ➢ Where are coniferous trees found? • In mountainous areas ➢ Does snow fall in those regions? • Yes ➢ Now tell me why the conifers are conical in structure. • So that snow can slide down.

Topic/Unit	*Main Points*	*Components*	*Procedural Details*
		Refocusing and redirecting	The teacher uses questions to refocus and redirect specific points such as: ➢ Why evergreen trees are found in the foothill regions whereas alpine vegetation is found in the mountains? • Due to differences in soil types ➢ Is there any relation between temperature and altitude? • With the increase in altitude temperature decreases ➢ Then tell me the reason why natural vegetation will differ. • Due to the difference in temperature

Skill of Use of Teaching Aids

Topic/Unit	*Main Points*	*Components*	*Procedural Details*
Natural Vegetation of India	• Concept and Importance of Natural Vegetation • Types of Natural Vegetation • Measures for Conservation of Forests • Social-Forestry and Agro-Forestry	Relevant to the topic	The teacher adds that aids should be relevant to the topic such as: • A short movie clip demonstrating the mangrove forests of Sunderban. • A pictorial chart on 'Different Types of Natural Vegetation and their Adaptive Mechanism' • A pictorial chart on 'Some important species of different types of natural vegetation'.
		Appropriate to the pupil's level	The chart must be colourful and appropriate to the student's level of maturity.
		Creating interest and thinking	The aids must initiate faster thinking in the students, such as: The students will see the pictures of the adaptive features of different types of natural vegetation and will learn the importance of these adaptive features as they read through the chapter.
		Proper display	The charts will be displayed properly so that the students can see them properly and clearly. The letters in the charts should be bold and they will be written in bright colours against a light backdrop so that it is visible to everyone. The chart should be hung up high so that it is visible from the last bench.

Topic/Unit	*Main Points*	*Components*	*Procedural Details*
		Appropriate use	The teaching aids should be used appropriately. For Example- The movie clip demonstrating the mangrove forests of Sunderban should be shown at the commencement of the lesson to secure the attention of the students. • The chart showing the pictures of different types of natural vegetation and their adaptive mechanism should be displayed throughout so that the student can follow the chart and can easily understand when the teacher explains each type of natural vegetation.

Clips of Blackboard Work done by the Teacher

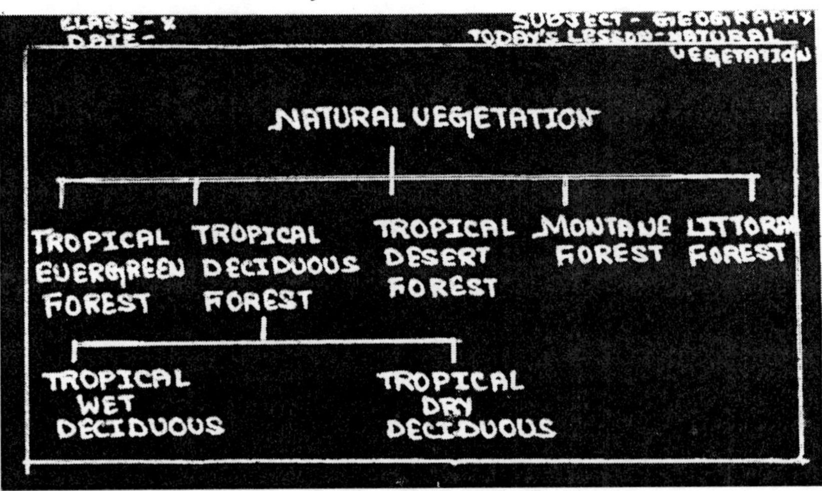

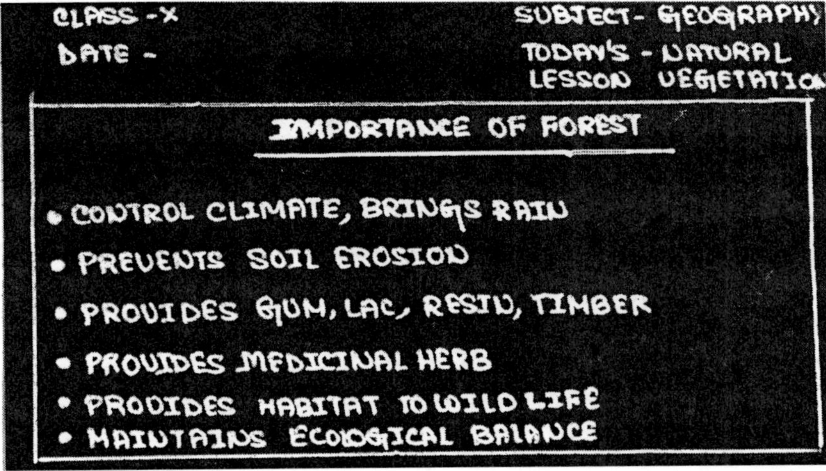

PICTORIAL CHART: TYPES OF NATURAL VEGETATION AND THEIR ADAPTIVE MECHANISMS

TROPICAL EVERGREEN FOREST

LAYERS OF TROPICAL EVERGREEN FORESTS

HUGE BUTTRESSED TRUNK OF TROPICAL EVERGREEN TREES

CONIFEROUS FOREST

NEEDLE SHAPED LEAVES OF CONIFEROUS TREES

PICTORIAL CHART: TYPES OF NATURAL VEGETATION AND THEIR ADAPTIVE MECHANISMS

TROPICAL DESERT FOREST

ADAPTIVE MECHANISM OF THE XEROPHYTES

TROPICAL DECIDUOUS FOREST

TROPICAL DECIDUOUS FORESTS SHED THEIR LEAVES

MANGROVE FORESTS WITH ANCHORING STEMS

PNEUMATOPHORES OF THE MANGROVES

Skill of Reinforcement

Topic/Unit	*Main Points*	*Components*	*Procedural Details*
Natural Vegetation of India	• Concept and Importance of Natural Vegetation • Types of Natural Vegetation • Measures for Conservation of Forests • Social-Forestry and Agro-Forestry	Positive reinforcement; no negative reinforcement	The teacher should always use positive reinforcement like good, correct, fine, and well done when the student gives the correct answer. This type of behaviour from the teacher increases the self-esteem of the students and motivates them to learn. For example: ➢ Where social forestry is best practiced? • Kerala ➢ Very good The teacher should never use negative reinforcement like bad, very bad, or wrong as it demotivates the students.
		Wrong response weakened or broken	The teacher breaks or weakens the wrong responses of the students in a way that doesn't demotivate the students but redirects them toward the correct responses. For example: ➢ Why shifting cultivation has been banned in the north-western states of India? • Because it leads to land degradation ➢ There is a different reason behind it. Think a bit more and answer. Didn't shifting cultivation destroy the forests? • Shifting cultivation leads to deforestation and for the conservation of the forests it has been banned.
		Providing lead or seeking further information	Instead of scolding the students, the teacher provides clues to the students in case they can't answer a question or give a complete answer. Such as: ➢ Why do the xerophytes have waxy stems? • To limit the loss of water through transpiration. ➢ Think a bit more and give the complete answer. Can you tell me where the xerophytes are found? • In the desert. ➢ Is there a scarcity of water in the desert? • Yes ➢ Now tell me why do the xerophytes develop waxy stems? • In the deserts due to lack of rainfall there is a scarcity of water; hence to limit the loss of water through transpiration the xerophytes develop waxy stems.

Topic/Unit	*Main Points*	*Components*	*Procedural Details*
		Positive non-verbal reinforcement	The teacher should use positive non-verbal reinforcement to motivate the students when the student gives the correct answer to a question. The teacher should appreciate the answer by smiling and nodding.
		Writing pupil's response on the board	The teacher should write the correct answer of the students on the board. This encourages the students and creates a feeling of acceptance among them.

Skill of Illustration

Topic/Unit	*Main Points*	*Components*	*Procedural Details*
Natural Vegetation of India	• Concept and Importance of Natural Vegetation • Types of Natural Vegetation • Measures for Conservation of Forests • Social-Forestry and Agro-Forestry	Relevant examples	The teacher should give relevant examples while explaining a particular topic such as: While explaining the coniferous forests, the teacher will give examples of some trees of coniferous forests (like Pine, Fir, and Spruce) and will show the picture of those trees displayed on the pictorial chart.
		Clear concept	Good examples help in building concepts. The teacher should give relevant examples while explaining a particular topic so that the concept becomes vividly clear to the students. For example, while explaining Tropical Evergreen Forests, the teacher will give an example of the film 'Sabuj Deeper Raja' where the students can see dense evergreen forests.
		Simple and easy	The teacher should give simple and easy examples while explaining a particular topic. The examples should be appropriate to the student's level of maturity.
		Student's interaction	The teacher should encourage student interaction so as to ensure that all the students actively participate in the classroom discussion. For example: After explaining the characteristics of tropical deciduous trees the teacher asks a few questions. ➢ Give an example of a tree that sheds its leaves during winter. • Palash ➢ Name some other trees of tropical deciduous forest. • Teak, Sal
		Correlates with the topic	The teacher should give such examples that correlate with the topic such as: While explaining the conservation of forests the teacher may give example of 'Chipko Movement' conducted by Sunderlal Bahuguna in Uttarakhand where the women embrace the trees to prevent it from being cut down.

Skill of Closure

Topic/Unit	*Main Points*	*Components*	*Procedural Details*
Natural Vegetation of India	• Concept and Importance of Natural • Vegetation • Types of Natural Vegetation • Measures for Conservation of Forests • Social Forestry and Agro-Forestry	a) **Summarization** – When the lesson was introduced, the main points in the lesson were specified. At the end of the lesson, the teacher concludes the lesson with a clear indication that the same points have been covered and a summary of the lesson has to be prepared accordingly.	The teacher will specify the major points of importance, types (by pointing to the learning aid), and measures of conservation of natural vegetation on the Blackboard (writing only *keywords*) in summary. The teacher also distinguishes between Social and Agroforestry prevalent in India for better understandability.
		Linked with the future learning – Indication is given about what may be the implication of the lesson in the next lesson.	The teacher discusses the correlation between natural vegetation and settlement density and also specifies the agricultural and industrial practices associated with it.
		Link with real-life activities – Whether the lesson has any bearing upon the real-life or out-of-school experience and if there is any such link it is to be exactly pointed out at the end of the lesson.	They would be aware 1. To save paper as it's a product of forest 2. To sensitize at school and at home about the importance of trees through programs 3. To be aware of the contemporary issues regarding forests and their fauna, forest products 4. To understand the interdependence of forest and community
		Evaluation – Whether immediate evaluation has been done during closure. This is necessary to assess the success or failure of the lesson.	The teacher asks a few hinge-based questions to check whether the students have consolidated and assimilated the content and are now ready to move on to the next.
		Adequacy of assignment – Whether assignments are given appropriately and whether assignments pertain to the domains of objectives other than cognitive, particularly to the affective domain need to be seen.	The teacher can assign relevant homework on written or project-based activities on an individual basis. Pertaining to the topic taught, group activities can also be conducted and completed within the class hour.